Solid-State Additive Manufacturing

The text focuses on discussing the solid-state deformation behavior of materials in additive manufacturing processes. It highlights the process optimization and bonding of different layers during layer-by-layer deposition of different materials in Solid-State. It covers the design, process, and advancement of solid-state additive manufacturing methods.

- Covers the fundamentals of materials processing, including the stress–strain–temperature correlation in solid-state processing and manufacturing.
- Discusses solid-state additive manufacturing methods, and optimization strategies for the fabrication of additive manufacturing products.
- Showcases the mechanisms associated with improvement in mechanical properties of Solid-State additive manufacturing products.
- Provides a comprehensive discussion on microstructural stability and homogeneity in mechanical properties.
- Presents hybrid solid-state process for fabrication of multilayer components and composite materials.
- Provides a detailed review of laser-based post-processing techniques

The text focuses on the Solid-State additive manufacturing techniques for the fabrication of industrially relevant products. It gives in-depth information on the fundamental aspects, hybridization of the processes, fabrication of different materials, improvement in product performance, and Internet of Things enable manufacturing. The text covers crucial topics, including hybrid Solid-State additive manufacturing, cold spray additive manufacturing, online defect detection of products, and post-processing of additively manufactured components. These subjects are significant in advancing additive manufacturing technology and ensuring the quality and efficiency of the produced components. It will serve as an ideal reference text for senior undergraduate and graduate students, and researchers in fields, including mechanical engineering, aerospace engineering, manufacturing engineering, and production engineering.

Solid-State Additive Manufacturing

Edited by
Amlan Kar
Zafar Alam

CRC Press
Taylor & Francis Group
Boca Raton London New York

CRC Press is an imprint of the
Taylor & Francis Group, an **informa** business

First edition published 2024
by CRC Press
2385 Executive Center Drive, Suite 320, Boca Raton FL 33431

and by CRC Press
4 Park Square, Milton Park, Abingdon, Oxon, OX14 4RN

CRC Press is an imprint of Taylor & Francis Group, LLC

© 2024 selection and editorial matter, Amlan Kar and Zafar Alam, individual chapters,

ISBN: 978-1-032-39274-5 (hbk)
ISBN: 978-1-032-61601-8 (pbk)
ISBN: 978-1-032-61602-5 (ebk)

DOI: 10.1201/9781032616025

Typeset in Sabon
by SPi Technologies India Pvt Ltd (Straive)

Contents

Preface

The book *Solid-State Additive Manufacturing* describes various Solid-State additive manufacturing (AM) techniques for the fabrication of AM parts/products. It provides a comprehensive overview of the fundamental aspects of Solid-State AM, including the hybridization of processes, the fabrication of different materials, the improvement in product performance, and Internet of Things (IoT)-enabled manufacturing. Conventional AM products suffer from low strength and toughness due to the melting of materials, induced residual stress, and microstructural heterogeneity. These products also exhibit low mechanical properties owing to metallurgical bonding and are therefore not directly useful in industrial applications, such as in aviation and medical environments. Solid-State AM operations overcome these limitations by imparting high mechanical properties and high-strength bonding between layers while manufacturing multi-component and complex products. This eliminates the need for additional post-processing to improve the mechanical properties of AM products developed by conventional liquid phase AM. The aim of this book is to bring together the current research archives from various sources to create a platform that provides readers with all the Solid-State AM techniques, together with the latest research technologies in these processes. The book is well organized to impart in-depth knowledge of different Solid-State AM parts to cater to the needs of a variety of industries.

The book contains eight chapters in total, starting with an Introduction and Solid-State Deformation Behavior of Materials in Additive Manufacturing (Chapter 1), Hybrid Additive Manufacturing using Cold Spraying and Friction Stir Processing (Chapter 2), Additive Friction Stir Deposition (Chapter 3), Cold Spray Additive Manufacturing (Chapter 4), Progress in Solid-State Additive Manufacturing of Composites (Chapter 5), Role and Capability of Internet of Things in Additive Manufacturing and its Application Regimes (Chapter 6), Assessment of Defects in Solid-State Additive Manufacturing through Conventional Methods and Sensor Based Data (Chapter 7), and Laser-based Post-processing Technologies for Additive Manufactured Parts (Chapter 8). The chapters are well organized to cover all aspects of Solid-State AM techniques, including basic deformation mechanism, physics involved in different AM processes, microstructure and mechanical properties of different materials fabricated by Solid-State AM process, and detail of equipment involved in these followed by limitations where applicable.

Every chapter is well supported by the related diagrams, figures, and actual pictures for better realization. The content of this book is intended to cater to the needs of academicians, engineers, researchers, practitioners, and post-graduates. The editors have very comprehensively brought together the current research archives from various sources to come up with a platform that has all the Solid-State AM techniques, together with the latest research technologies in these processes for the readers to benefit from. This easy manner of understanding provides the readers with an opportunity to achieve a deep understanding of these techniques. This book is an equal combined effort on the part of the co-editors, Dr. Amlan Kar and Dr. Zafar Alam. It includes novel and unique contributions from a wide range of authors drawn from across the globe.

We will happily acknowledge suggestions from the readers of this book to improve the content of this book and will make every attempt to incorporate any appropriate suggestions in the next edition.

Amlan Kar
Zafar Alam

Editors

Dr Amlan Kar is a Senior Scientist at the South Dakota School of Mines and Technology. He has a background in mechanical engineering and previously served as a faculty member at the Indian Institute of Technology (Indian School of Mines) Dhanbad in India. He earned his B.Tech. in Mechanical Engineering from Jalpaiguri Government Engineering College and went on to pursue an M.Tech. in Metal Forming at the Indian Institute of Technology Madras. He was fascinated by the structure–property correlation of materials during plastic deformation, which led him to earn his PhD in Dissimilar Friction Stir Welding from the Indian Institute of Science (IISc) Bangalore. Dr. Kar has worked on various industrially relevant projects, including the development of composite materials using Friction Stir Processing and the incremental sheet metal forming of titanium. He has also contributed to the field of Friction Stir Welding, particularly in the joining of dissimilar light alloys. His expertise extends to dissimilar Friction Stir Lap welding of aluminum to steel using an adjustable tool, conducted at JWRI, Osaka University. Currently, Dr. Kar focuses his research on dissimilar welding techniques, mechanical joining, additive manufacturing, and other related areas. His goal is to improve the mechanical properties of materials across different temperatures through a deeper understanding of their mechanical behavior. Dr. Kar is an accomplished academician and researcher. He has published numerous papers in renowned international journals and conferences. He has received prestigious awards, including the "IIM SJEF Fellowships," and serves as a reviewer for esteemed journals. Dr. Kar is actively involved in professional committees such as TMS and the Indian Society for Applied Mechanics, leveraging his expertise in welding, metal forming, additive manufacturing, materials characterization, and structure–property–performance correlation.

Dr Zafar Alam is an Assistant Professor at the Department of Mechanical Engineering, Indian Institute of Technology (Indian School of Mines) Dhanbad, India. Prior to joining IIT (ISM) Dhanbad, he served for two years as an Assistant Professor of Mechanical Engineering at Zakir Husain College of Engineering and Technology, Aligarh Muslim University, India. He earned his B.Tech. in Mechanical Engineering from Jamia Millia Islamia, India in 2012. He earned his M.Tech. and PhD in Production Engineering from the Indian Institute of Technology Delhi, India in 2014 and 2019, respectively. As an academician and researcher, he has published both an authored and an edited book, several book chapters and more than two dozen research papers in peer-reviewed international journals and conferences. He also has to his credit six Indian patents and has received two international and three national awards, including the critically acclaimed Gandhian Young Technological Innovation (GYTI) award for his contributions in the field of research and innovation. His research interests include, but are not limited to, advanced finishing/polishing processes, non-conventional machining, additive manufacturing, industrial automation and motion control.

Contributors

Zafar Alam
Indian Institute of Technology
 (Indian School of Mines)
 Dhanbad
Dhanbad, India

Bhavesh Chaudhary
Indian Institute of Technology Indore
Indore, India

Xin Chu
Institute of New Materials
Guangdong Academy of Sciences
Guangzhou, China

Vipin Gopan
St. Thomas College of Engineering
 and Technology
Alappuzha, India

Jibo Huang
South China University of
 Technology
Guangzhou, China

Renzhong Huang
Hubei Chaozhuo Aviation
 Technology Co., Ltd.
Xiangyang, China

Faiz Iqbal
University of Lincoln
Lincoln, United Kingdom

Neelesh Kumar Jain
Indian Institute of Technology Indore
Indore, India

Amlan Kar
South Dakota School of Mines &
 Technology
Rapid City, South Dakota

Dilshad Ahmad Khan
National Institute of Technology
 Hamirpur
Hamirpur, India

Gaurav Kishor
Sardar Vallabhbhai National
 Institute of Technology Surat
Surat, India

Praveen Kumar
Mahaguru Institute of Technology
Alappuzha, India

Sachin Kumar
NamTech Institute
Gandhinagar, India

Haiming Lan
Hubei Chaozhuo Aviation
 Technology Co., Ltd.
Xiangyang, China

Shrushti Maheshwari
Indian Institute of Technology
 (Indian School of Mines) Dhanbad
Dhanbad, India

Raju Prasad Mahto
Sardar Vallabhbhai National
 Institute of Technology Surat
Surat, India

Krishna Kishore Mugada
Sardar Vallabhbhai National
  Institute of Technology Surat
Surat, India

Jayaprakash Murugesan
Indian Institute of Technology
  Indore
Indore, India

Valliyappan David Natarajan
College of Engineering
Universiti Teknologi MARA
Shah Alam, Malaysia

Mahesh Patel
Indian Institute of Technology
  Indore
Indore, India

Vivek Patel
University West
Sweden

Jibin T. Philip
Indian Institute of Science
Bengaluru, India

S. Sarath
St. Thomas College of Engineering
  and Technology
Alappuzha, India

Aditya Sharma
Dayalbagh Educational Institute
Agra, India

Ashish Siddharth
Indian Institute of Technology
  (Indian School of Mines)
  Dhanbad
Dhanbad, India

Kuldeep Singh
Indian Institute of Science
Bengaluru, India

Wen Sun
Hubei Chaozhuo Aviation
  Technology Co., Ltd
Xiangyang, China

Roshin Thomas Varughese
St. Thomas College of Engineering
  and Technology
Alappuzha, India

Chapter 1

Introduction and solid-state deformation behavior of materials in additive manufacturing

Amlan Kar
South Dakota School of Mines & Technology, Rapid City, SD, USA

1.1 INTRODUCTION

Additive manufacturing (AM) is a process of creating objects layer-by-layer, typically using computer-aided design (CAD) software to create a 3D model of the object and a 3D printer to produce it. This process is in contrast to subtractive manufacturing, which involves cutting or removing material from a larger piece until the desired shape is achieved. AM has many advantages, including the ability to produce complex geometries, reduce waste, and create customized products quickly and cost-effectively. It is used in a wide range of industries, from aerospace and automotive to healthcare and consumer goods. The blacksmith in ancient times was mainly concerned with the change in shape in any metallic component by metal forming operations, such as forging, rolling, extrusion, etc. The shape change in metallic components was achieved by thermomechanical processing mostly in the solid-state. Industrial products require a certain optimum property in terms of stress, strain, and other physical properties apart from dimensional accuracies. Optimum physical properties have a direct correlation with the microstructure of the components, which is dependent on the external thermomechanical parameters, such as temperature, strain, strain rate, deformation conditions, etc. The parameters are process-specific that convert the materials to design specific components with higher mechanical properties. The thermomechanical process was extensively used from ancient times, being involved in the development of personal ornaments and decorative items of bronze during the Bronze Age (3000 BC). In the ancient age, cast materials were subjected to temperature and plastic deformation to change the shape of materials to a useful component with improved mechanical properties. In modern times, temperature and plastic deformation have been employed in a controlled environment and with greater precision. Scientific advancement promotes the extensive use of optimum conditions to achieve a particular goal while processing a material under different thermomechanical conditions. In the last few decades, correlation between process and properties have been extensively studied. It has been accepted that microstructural evolution during thermomechanical process play a decisive role

DOI: 10.1201/9781032616025-1

in optimizing the process–property correlation. Thermomechanical process always performs under solid-state conditions, below the melting point of pure materials and the solidus temperature of alloys. The solid-state deformation behavior under thermomechanical processing can be described by both mechanical approaches and microstructural approaches. The mechanical approach is characterized by stress–strain behavior at different strain rates and temperatures. On the other hand, the microstructural approach describes the dislocation activity and corresponding strain activity in the grains, leading to a variation in mechanical properties. The aim of the chapter is to bring the two fundamental approaches closer together to map a direct correlation that could facilitate new advances in the field of solid-state additive manufacturing.

1.2 INTRODUCTION TO SOLID-STATE AM

Solid-state additive manufacturing (AM) is a category of metal AM techniques that use solid-state processes, as opposed to melting and solidification processes used in beam-based methods. These techniques are categorized based on the metallurgical bonding mechanism, the range of processible alloys, and the resulting microstructures. Deformation-based methods, such as additive friction stir AM, refine the microstructure through recrystallization, while sinter-based AM methods lead to grain growth. Isotropic microstructures and mechanical properties close to wrought properties can be achieved using Binder Jetting and additive friction stir AM methods. Ultrasonic additive manufacturing (UAM) is suitable for relatively soft austenitic steels or steels that undergo softening by austenitic transformation. However, high-hardness steels and nickel tend to stick to the sonotrode, limiting the range of processible alloys. Multi-material printing is possible using UAM because of its low processing temperatures. Solid-state metal AM technologies offer a wide range of capabilities, including the ability to process a large portfolio of metals and alloys, unique microstructures, and ease of multi-material manufacturing. The bonding mechanism is a critical aspect that determines the processible materials, resulting properties, and speed and efficiency of the process. Sinter-based AM is more prone to grain growth due to prolonged exposure to high temperatures during sintering. Additive friction stir AM and Binder Jetting are two solid-state AM methods capable of achieving isotropic bonding strength in the as-built state. Efforts towards developing new alloys and improving bonding quality to achieve good mechanical properties and isotropy will further expand the potential of solid-state additive manufacturing.

Solid-state additive manufacturing (AM) refers to a category of AM processes that create 3D parts by consolidating materials in the solid-state, without melting them. The following are some of the types of solid-state AM:

a) Sintering-based solid-state AM: In this process, a 3D-printed powder compact is consolidated by heating it uniformly, causing the particles to fuse together. This technique is often used for ceramics and metals.

b) Mechanical deformation (MD)-based techniques: This method uses mechanical energy to facilitate metallic bonding. Additive Friction Stir Deposition (AFSD), Ultrasonic Additive Manufacturing (UAM), and Cold Spray Additive Manufacturing (CSAM) are examples of MD-based techniques.

c) Material Extrusion-Based Additive Manufacturing (MEAM): MEAM is a process of material extrusion based on fused filament fabrication (FFF). This method is often used for polymers and composites.

d) Screen and Stencil 3D Printing (SPAM): This process involves forcing a paste or slurry of a material through a mesh or stencil to create a 3D part.

e) Binder Jetting: Binder Jetting is a process that uses a liquid binding agent to selectively bond powder particles together, creating a 3D part. This method is commonly used for ceramics and metals. An industrial print head is used to selectively apply the binding agent to a fine layer of powder material during the build process. This creates a solid and complex part.

f) Directed Energy Deposition (DED): It is a 3D printing process that uses a focused energy source, like a laser or electron beam, to melt and fuse metal powder or wire feedstock. This method is commonly used for repairing or adding material to existing parts. During the process, material is melted upon deposition, and further material is added layer by layer and solidifies, creating or repairing new features on the existing object. DED is a term that encompasses all the technologies involving semi-automated powder spraying and wire welding. This process offers advantages such as the ability to print large structures quickly and the ability to print functionally graded materials.

g) Sheet Lamination: Sheet Lamination is an additive manufacturing process that involves layer-by-layer bonding thin sheets of material together to form a single piece that is then cut into a 3D object. Paper, plastic, and metal sheets are common materials for this technique. The sheet lamination process is divided into three stages: paper-based lamination, composite-based lamination, and selective lamination. Paper-based lamination entails gluing layers of paper and foils together, which are then precisely cut to the desired geometry to create the final object.

h) Vat Photopolymerization: Vat Photopolymerization is a type of additive manufacturing (AM) process that creates 3D objects by selectively curing liquid resin through targeted light-activated polymerization. The process uses a vat of liquid photopolymer resin, and the model is built layer-by-layer. A UV light cures or hardens the resin where required, and a platform moves the object downwards after each new layer is cured. This technique is particularly suitable for producing high-resolution models and prototypes. Stereolithography, the first AM process to be patented and commercialized, is a type of vat photopolymerization.

Additive manufacturing (AM) for metals and alloys often utilizes mechanical deformation (MD)-based techniques that produce parts with high density and excellent mechanical properties, including strength, toughness, and ductility. These techniques can achieve higher densities compared to sintering-based techniques because the mechanical energy facilitates metallic bonding between the powder particles, resulting in fewer pores and voids in the final part. Moreover, MD-based techniques offer advantages over traditional manufacturing processes by enabling the production of complex geometries that are difficult or impossible to produce using conventional methods, with minimal waste and without any need for expensive tooling. There are several types of MD-based techniques available for solid-state additive manufacturing, including:

a) Additive Friction Stir Deposition (AFSD): This technique is a solid-state deposition manufacturing process that uses a rotating tool to mechanically bond layers of material together. This technique is based on the principles of friction-stir welding technology and can be classified into three modes: powder with a non-consumable tool, rod with a non-consumable tool, and consumable tool. AFSD is a low-temperature process that is capable of building, coating, joining, or repairing three-dimensional metal objects. It offers several advantages over traditional manufacturing methods, including the ability to produce complex geometries with high accuracy and precision, minimal material waste, and no need for expensive tooling.

b) Ultrasonic Additive Manufacturing (UAM): This is an innovative method of solid-state deposition manufacturing that involves the ultrasonic welding of metal foils or powders layer by layer. It is a low-temperature process that allows for the efficient production of complex work, including parts with embedded components and even parts made from dissimilar metals. UAM builds metal workpieces by fusing and stacking metal strips in a continuous fashion, similar to sheet lamination in additive manufacturing. The process works by scrubbing metal foils together with ultrasonic vibrations under pressure, and melting is not the formation mechanism. Instead, metals are joined in the solid-state via the disruption of surface oxide films between the metals. Computer numerical control (CNC) contour milling is used to introduce internal features and add detail to the metal part during the additive stage of the process. UAM is capable of joining multiple metal types together, including dissimilar metals, with minimal intermetallic formation and allows the embedment of temperature-sensitive materials at relatively low temperatures, typically less than 50% of the metal matrix melting temperature [1, 2].

c) Cold Spray Additive Manufacturing (CSAM): Metal powders are deposited onto a substrate via a plastic deformation process in Cold Spray Additive Manufacturing (CSAM) by accelerating the powders to high velocities. This technology is used to build parts by combining multiple layers of continuous coatings on a substrate. CSAM is a subset of the cold spraying process, which is also used to modify existing components or create freestanding parts. Cold spraying is a coating deposition technique that employs high-velocity gas streams to accelerate fine powder particles.

1.3 CORRELATION BETWEEN MECHANICAL DEFORMATION (MD)-BASED TECHNIQUES AND A STRESS-STRAIN DIAGRAM

Mechanical deformation (MD)-based techniques, including friction stir processing (FSP) and ultrasonic additive manufacturing (UAM), utilize mechanical energy to facilitate metallic bonding, and can affect a material's microstructure, leading to changes in its mechanical properties. The stress-strain diagram is a useful tool to graphically represent a material's mechanical behavior under an applied load. It demonstrates the relationship between stress (force per unit area) and strain (resulting deformation) as the material is loaded and unloaded. The stress–strain diagram provides valuable information about a material's strength, stiffness, and ductility. MD-based techniques such as FSP and UAM can cause plastic deformation due to high shear forces and strains, leading to changes in the material's microstructure, such as grain refinement and recrystallization. These changes can affect key parameters like yield strength, ultimate tensile strength, and ductility, which are typically characterized by the stress-strain diagram. Therefore, the correlation between MD-based techniques and stress-strain diagram lies in the fact that the former can modify the microstructure and mechanical properties of a material, which can be subsequently measured and represented by the latter. The stress-strain diagram can thus be used to evaluate the effect of MD-based techniques on a material's mechanical behavior, and to optimize processing parameters to achieve the desired mechanical properties.

1.4 STRAIN–STRESS–TEMPERATURE CORRELATION

The strain-stress-temperature correlation in solid-state deformation of materials refers to how a material's properties change as it is subjected to different levels of stress, strain, and temperature. In general, as temperature increases, a material will become more pliable and have a lower yield strength (the point at which it begins to permanently deform). This is known as thermal softening. As stress increases, a material will experience an increase in strain (deformation) until it reaches its yield point, at which point it will experience plastic deformation and no longer return to its original shape. Additionally, different types of materials have unique behaviors. For example, some materials will exhibit a linear relationship between stress and strain up to a certain point and then will exhibit non-linear behavior. Additionally, some materials will exhibit a change in their behavior at certain temperatures, becoming either more brittle or more ductile. This correlation can be determined by experiments, and the results can be used to predict the behavior of the material under different loading conditions and temperatures, which is important in the design of engineering structures and components. Solid-state deformation behavior of materials follows the plasticity of materials at strains where materials do not follow Hooke's Law. Plastic deformation is not a reversible process and dimension samples only depends on the initial and final state of stress, strain and temperature. The analysis of deformation behavior is required in mathematical formulation. Deformation mechanics also vary with the mechanics of the plastic deformation of metals, which is directly dependent on crystal imperfection and dislocation activity inside the microstructure of deforming material. The flow curve reveals the actual characteristics of the material under deformation conditions. The flow curve is represented by a true stress vs. true strain plot at different temperatures and strain rates.

1.4.1 True stress-strain curve determination

True stress-strain curve determination is the process of measuring the stress and strain of a material under load. In practice, engineering stress-strain curves provide basic information about the strength of materials. This is typically done by testing a sample of the material in a tensile testing machine, which applies a controlled load to the sample while measuring the corresponding deformation. The results of the test are then plotted on a graph, with stress on the y-axis and strain on the x-axis. This graph, known as the engineering stress-strain curve, can be used to determine the material's properties, such as its yield strength, ultimate strength, and ductility. The following equations are used to calculate engineering stress (σ_e) and engineering strain (ε_e) based on observation during tensile test;

$$\sigma_e = \frac{P}{A_o} \tag{1.1}$$

$$\varepsilon_e = \frac{L - L_o}{L_o} = \frac{\Delta l}{L_o} \tag{1.2}$$

In both of the above equations, force and elongation have been divided by an initial dimension of the specimen, and hence load vs elongation would have an identical curve.

Even though the engineering stress-strain curve is simple to use and extensively used in industrial applications, it does not provide a true indication of the deformation behavior of material because it is based on the initial dimension of specimen that changes continuously with deformation. Therefore, a measure of true stress and true strain, which is important for metalworking operations such as severe plastic deformation, is desirable. The following equations are used to calculate true stress (σ) and true strain (ε) based on instantaneous dimensions of a tensile specimen.

$$\sigma = \frac{P}{A} = \sigma_e\left(\varepsilon_e + 1\right) \tag{1.3}$$

$$\varepsilon = \int_{L_o}^{L}\left(\frac{\Delta l}{L_o}\right) = \ln\frac{L}{L_o} = \ln\left(\varepsilon_e + 1\right) \tag{1.4}$$

Mathematically, it is convenient to use true stress and true strain as total true strain in a multiple-step deformation process is a sum of the true incremental strains. Below the elastic strain, most metallic materials do not have any difference in true and engineering stress and strain values. Therefore, the use of true stress and strain is important for plastic deformation, such as solid-state additive manufacturing and metalworking operations.

1.4.2 Difference between true stress and flow stress

True stress is the stress calculated using the instantaneous cross-sectional area of a material, while flow stress is the stress calculated using the original, unchanging cross-sectional area of a material. In other words, true stress takes into account any changes in the material's dimensions due to deformation, while flow stress assumes that the dimensions remain constant. This means that true stress will increase as a material is deformed and its dimensions change, while flow stress will remain constant. In constituent equations, true stress and flow stress are often used interchangeably. As discussed earlier, true stress, as the definition suggests, it is measured by instantaneous load/instantaneous area. Whereas the flow stress is the stress that must be applied to cause a material to deform at a constant strain rate in its plastic range. From these two definitions, it can be understood that true stress is measured from the start to the end of the deformation process whereas flow stress is measured during plastic deformation only, i.e. by removing elastic parts from the true stress-strain curve we get the flow curve. There are slight differences between the true and flow stress. Even though true stress and flow stress are relatively close values, they are not identical. In plotting and calculating flow stress, the true stress–true strain curve is modified to remove the elastic effect for perfect plastic flow analysis. Therefore, to measure strain–rate–sensitivity exponent "m", the slope is measured using flow stress-flow strain instead of true stress–true strain.

Many attempts have been made to fit the flow stress–strain curve mathematically. The common mathematical equation is called the power law of flow curve that is expressed as follows:

$$\sigma = K\varepsilon^n \tag{1.5}$$

Where, K is a constant at $\varepsilon = 0$ and n is the strain-hardening exponent that is measured by the slope of a log-log plot of Equation 1.5. This is ideally suitable in the region of uniform straining, as shown in Figure 1.1(b). The mathematical representation represented here looks simple, but it is difficult to use in theory of plasticity due to mathematical complexity. Therefore, it is very common practice to devise an idealized flow curve that can be used in the theory of plasticity without deviation from actual physical reality. Three types of idealized flow curves are commonly used: (a) rigid ideal plastic material, (b) ideal plastic material with elastic behavior and (c) piecewise liner material with linear strain hardening (Figure 1.2). Specific behavior of these three types of ideal materials is tabulated in Table 1.1.

1.4.3 Effect of temperature on flow stress

The effect of temperature on flow stress, also known as the temperature dependence of flow stress, can vary depending on the material and the

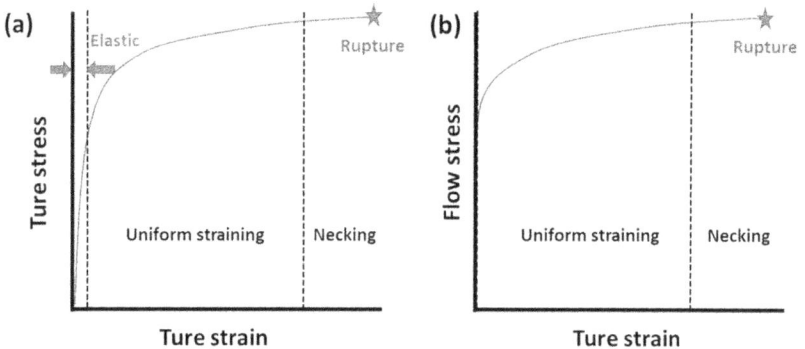

Figure 1.1 Schematic representation of (a) true stress-true strain and (b) flow stress curve.

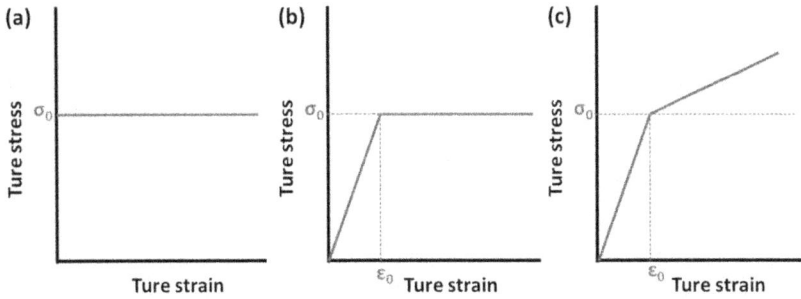

Figure 1.2 Schematic representation of idealized flow curves showing (a) rigid ideal plastic material, (b) ideal plastic material with elastic behavior, and (c) piecewise linear materials with linear strain hardening.

Table 1.1 Specific characteristics of idealized materials suitable for mathematical representation

Information	Rigid ideal plastic material	Ideal plastic material with elastic behavior	Piecewise linear materials with linear strain hardening
Rigidity; zero elastic strain	Yes	No	No
Elastic deformation	No	Yes	Yes
Strain hardening rate	Zero	Zero	Non-zero
Mathematic complexity	Very easy	Moderate	Complex
Example	Ductile materials with heavy cold worked	Plain carbon steel having high yield point elongation	Almost all realistic material

specific conditions of the test. In general, an increase in temperature can lead to a decrease in flow stress. This is because as the temperature increases, the atoms in the material have more kinetic energy and are able to move more easily, making it easier for the material to deform. However, the effect of temperature on flow stress can also depend on the initial microstructure and the rate of deformation. For example, in some materials, an increase in temperature can lead to an increase in flow stress at high strain rates due to dynamic recovery or dynamic recrystallization. Additionally, some materials exhibit a peak or a shoulder in the flow stress-temperature curve, which is known as the "peak temperature" or "shoulder temperature". This is an indication of a change in the deformation mechanism, usually it is caused by the onset of dynamic recovery. In summary, the effect of temperature on flow stress is complex and can vary depending on the material, the initial microstructure, and the rate of deformation. Solid-state deformation at low homologous temperature below $0.4T_m$, flow stress is independent of strain rate. The flow curve then follows the power law:

$$\sigma = K\varepsilon^n \tag{1.6}$$

The exponent, n, plays an important role in forming operation, and it is known as the strain-hardening exponent. It can be calculated by using the following equation,

$$n = \frac{d(\ln\sigma)}{d(\ln\varepsilon)} \tag{1.7}$$

This controls the amount of uniform plastic deformation materials can undergo without strain localization that leads to sample failure in universal tensile testing. If the onset of necking occurs at the maximum load, the amount of uniform plastic strain accrues the same value as the strain-hardening exponent:

$$\sigma = n \tag{1.8}$$

This limits the amount of uniform plastic deformation that can be applied in a room temperature deformation during additive manufacturing. For metallic materials, there is an increase in the load-carrying capacity of the specimen with the increase in deformation and/or n value. In solid-state deformation at high temperatures above $0.4T_m$, strain rate significantly influence the flow behavior of materials. Flow stress increases with the strain rate, as shown in Figure 1.3(b). The flow equation and strain rate sensitivity are defined as follow:

$$\sigma = C\dot{\varepsilon}^m \tag{1.9}$$

$$m = \frac{d(\ln\sigma)}{d(\ln\dot{\varepsilon})} \tag{1.10}$$

In which the value of m varies within the range 0.05–0.5 [3].

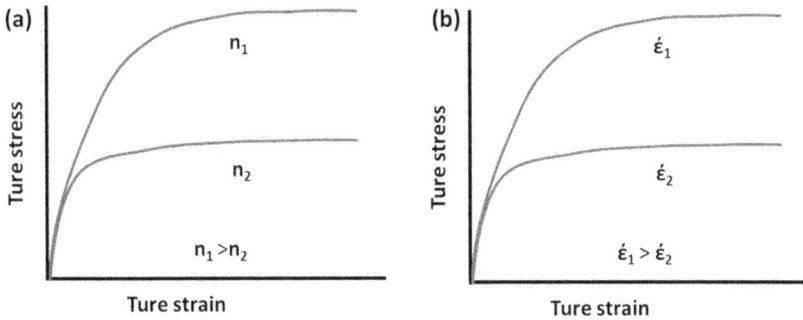

Figure 1.3 Schematic depiction of flow curve at (a) low temperature and (b) high temperature.

1.4.4 Correlation of stress, strain and temperature

In solid-state deformation, the relationship between stress, strain, and temperature can be described by the constitutive equation of the material. The constitutive equation relates the stress, strain, and temperature of a material at a given point in time, taking into account the material's physical properties and the external forces acting on it. The relationship between stress and strain is typically described by the material's stress-strain curve, which can change as the temperature of the material changes. In solid-state additive manufacturing, the relationship between stress, strain, and temperature is more complex due to the addition of material to build up the final product. The temperature of the material can vary widely during the process, and the addition of new material can also affect the stress and strain in the surrounding material. The solid-state deformation of materials in solid-state additive manufacturing takes place under hot-working conditions. At a higher homologous temperature beyond $0.4T_m$, plastic deformation is strongly influenced by the processing temperature and other related thermally activated processes. Then the flow stress of materials under deformation varies with temperature and strain rate. At $T > 0.4T_m$, dislocations in the material continuously rearrange themselves and thereafter are annihilated by climb and cross-slip mechanisms. Atomic diffusion at a higher homologous temperature assists dislocation movement and gives rise to dynamic recovery and recrystallization during solid-state deformation. Ghosh et al. [4], developed a set of constitute relations that descripted the hot deformation behavior of materials. A combined generalized relation, a unified model of existing various physical processes, illustrates the relation among flow stress, temperature and strain rate in dynamic process such as solid-state deformation. At a fixed strain, flow stress is a function of temperature (T) and strain rate ($\dot{\varepsilon}$) as described by a constituent relationship.

$$\sigma = f(Z) = f\left(\dot{\varepsilon}e^{\left(\frac{Q}{RT}\right)}\right)$$

(1.11)

Where, Q is an activation energy for plastic flow. Values of Q varies in the range of 150–200 kJ/mol for aluminum alloys, 280 kJ/mol for ferritic steel and 300 kJ/mol for austenitic stainless steel. Equation 1.11 is a useful constitutive relation correlative flow stress, strain rate and temperature under hot working conditions. Values of Q increase with alloying addition as there is an increase in resistance to dislocation movement. For low-or-medium stacking fault energy of face-centred cubic (FCC) materials (Example: austenitic stainless steel), the cross-slip of dislocation is hindered, and the rate of dynamic recovery is very slow; flow stress increase significantly with strain and strain rate at a particular temperature. Then accumulated dislocation density reaches a critical strain for dynamic recrystallization during solid-state deformation. The possible dynamic recrystallization reduces the flow stress.

The term 'Z' is called Zener–Holloman parameters that varies with flow stress and strain rate.

$$\sigma = Z = \dot{\varepsilon}e^{\left(\frac{Q}{RT}\right)}$$

(1.12)

At low Z value (Z < 26 for aluminum), deformation become viscous and a constant flow stress with strain rate can be obtained. This defines creep phenomena where an increase in stress leads to continuous shape change. With the increase in Z value, flow stress varies with strain and strain rate along with temperature. Seller et al. [5] developed a new and modified constitutive correlation for thermomechanical deformation processes as follow:

$$\dot{\varepsilon} = A\left(\sinh(\alpha\sigma)^x e^{\left(\frac{-Q}{RT}\right)}\right)$$

(1.13)

Where A, α, and x are constants. At lower stress ($\alpha\sigma < 1$) deformation process and high homologous temperatures, Equation 1.6 is converted to a power relation (creep type of behavior)

$$\dot{\varepsilon} = A\sigma^x e^{\left(\frac{-Q}{RT}\right)}$$

(1.14)

At lower stress ($\alpha\sigma > 1.2$) deformation process and high homologous temperatures, Equation 1.14 is converted into an exponential relation,

$$\dot{\varepsilon} = \exp(y\sigma)e^{\left(\frac{-Q}{RT}\right)}$$ (1.15)

Where the 'y' is a constant that is related by $y = \alpha x$. In all the above-mentioned equations, a particular constant can be determined from a log-log plot. For example, activation energy, Q, is generally measured from a plot of log $\dot{\varepsilon}$ vs log σ using Equation 1.15.

To describe the flow stress at different flow regimes of having a different Z value due to a variation in Q, empirically there are two possible approaches:

a) Voce law; $Z = f(\sigma)$ [6]

$$\sigma = Z = \sigma + (\sigma - \sigma)\left[1 - \exp\left(\frac{-\varepsilon}{\varepsilon_0}\right)\right]^z$$ (1.16)

Where ε_0 and z are transition deformation and exponent constant, respectively.

b) A modified Kocks-Mecking approach uses microscopic information, such as dislocation density. It is used to characterize the thermally activated plastic deformation to account for nonlinear strain hardening [7].

Solid-state deformation processes are an effective means of refining the grain size of different alloys [8–11]. A refined grain size in the range of 0.5–10 μm due to the dynamic recrystallization of aluminum alloys has been widely reported [12–14]. The processes employed a different level of fundamental parameters, such as stress, strain, strain rate and temperature, for grain refinement. Hence, there is a need of a simple correlation between these fundamental processing parameters and grain size. The Zenner Holloman parameter could influence and predict the resulting grain size [15]. For example, a simple relationship for friction stir processing (FSP), tensile test and the warm extrusion of AZ31 alloy was established by Chang et al. [16] as follows:

$$\ln d = m - n \ln Z$$ (1.17)

Where d is grain size, and m and n are constants. This is a straight-line equation and it confirms that grain size follows similar trends irrespective of the processing path. In this regard, Sato et al. [17] correlated the grain size with processing temperature during FSP of aluminum-6063 alloys. The correlated expression was

$$\ln d = \ln \frac{At}{2} - \ln \frac{Q}{2RT} \qquad (1.18)$$

Where A is a constant and t is time. Strain rate was calculated using the time t. Hence, the same equation can be used to study the fluence of strain rate and temperature on resulting grain size.

1.5 THE EFFECT OF TEMPERATURE IN SOLID-STATE DEFORMATION

Temperature plays a crucial role in both solid-state deformation and solid-state additive manufacturing processes. In general, increasing the temperature can make the material easier to deform or manipulate, but there is a limit to how high the temperature can be before the material loses its strength or begins to experience incipient melting in the case of alloys. In solid-state deformation, temperature can affect the stress and strain in the material, as well as the material's microstructure and physical properties. At higher temperatures, the material may experience more plastic deformation, allowing it to be formed into more complex shapes. However, at extremely high temperatures, the material may experience recrystallization, where new grains form and the microstructure of the material is altered, both of which can affect the material's strength and ductility. In solid-state additive manufacturing, temperature is critical for controlling the properties of the final product. The addition of new material requires heating to a specific temperature to allow for fusion with the surrounding material. The temperature can affect the quality of the final product, such as its density, porosity, and mechanical properties. If the temperature is too low, the material may not fuse properly, resulting in a weak or defective final product. If the temperature is too high, the material may melt or experience distortion or warping.

Solid-state-deformation can be classified as either hot or cold deformation. In the hot deformation process, recovery and deformation occurs simultaneously. Rapid strain hardening activates the recrystallization process and, hence, distorted grains are replaced by new strain-free grains. Recovery and recrystallization facilitate the additional plastic deformation of materials. Repeated solid-state deformation of materials during additive manufacturing along the build direction is possible during hot working conditions. By contrast, due to the slow atomic diffusion at cold temperatures, recovery is not an active mechanism in cold deformation during additive manufacturing. Atomic diffusion during subsequent passes, could be sufficient to activate the recovery process. However, strain hardening and corresponding accumulation of residual strain at the interface of two layers is not get eliminated. Therefore, the interface is the weak zone due to the accumulation of strain, which is prone to fracture. For this reason, the hot working of materials, especially alloys, is always preferable in order to activate recovery and recrystallization. Peak temperature and thermal exposure duration play an important role in microstructure evolution in deformed structures. Temperature evolution varies with: i) the pre-heat condition or initial temperature, ii) plastic deformation, iii) the frictional condition between the tool and the material and iv) internal and external heat transfer conditions. Temperature increases due to the deformation of materials can be calculated using the following equation:

$$T_d = \frac{W_p}{\rho c H} = \frac{\sigma \varepsilon \beta}{\rho c H} \qquad (1.19)$$

Where,

W_p = work per volume due to plastic deformation
P = density of workpiece material,
c = specific heat of workpiece material,
H = mechanical equivalent of heat,
β = the fraction of deformation work converted to heat
σ = average flow stress,
ε = average strain

Similarly, temperature rise due to interfacial friction can be calculated using the following equation:

$$T_f = \frac{\mu p v A \Delta t}{\rho c V H} \qquad (1.20)$$

Where,

μ = work per volume due to plastic deformation
P = interfacial coefficient of friction,
P = normal stress,
v = interfacial velocity,
A = interfacial surface area,
Δt = considerable time interval,
V = volume of material subjected to temperature rise.

Considering the heat transfer, the average instantaneous temperature at the interface is given by the following equation [3];

$$T = T_1 + (T_0 - T_1)\left(\frac{-ht}{\rho c \varphi}\right) \qquad (1.21)$$

Where,

h = heat transfer coefficient between material and die,

φ = thickness of materials between dies.

Taking into all the above-mentioned equations, the final temperature of deforming materials can be postulated as follows,

$$T_{dm} = T_d + T_f + T \tag{1.22}$$

1.6 WORK HARDENING DURING SOLID-STATE DEFORMATION

Solid-state additive manufacturing (SSAM) can induce work hardening in the deposited material due to the severe plastic deformation that occurs during the process. The repeated deposition of material layers, along with the localized heating and cooling, can cause the material to experience cyclic loading, leading to work hardening. The extent of work hardening depends on various factors, including the material being used, the process parameters, and the deformation rate. During the process of SSAM, the repeated deformation of the material can cause dislocations to accumulate, leading to an increase in the material's strength and hardness. The degree of work hardening can be quantified by measuring the increase in the material's yield strength or hardness after deformation. Work hardening can make the material more resistant to subsequent deformation, which can prove beneficial for certain applications. However, excessive work hardening can also lead to cracking or other defects in the deposited material. It can also affect the microstructure and other material properties, such as ductility and toughness. Therefore, it is important to optimize the SSAM process parameters to control the degree of work hardening and ensure that the final product has the desired properties. Post-processing treatments, such as annealing, can also affect the degree of work hardening in the SSAM material. Annealing can help to relieve the accumulated internal stresses and reduce the dislocation density, leading to a decrease in the material's strength and hardness. The optimization of the SSAM process parameters and post-processing treatments can help to achieve the desired balance between strength, ductility, and other material properties. Work hardening can be demarcated based on homologous temperature.

1.6.1 Work hardening at low temperature ($T < 0.4T_m$)

Thermally activated processes do not play a significant role in work hardening at low temperatures, defined as where T is less than 0.4 times the melting temperature (T_m) of the material. At these low temperatures, the plastic deformation of the material is primarily driven by dislocation activity along crystallographic planes under the application of external stress. Work hardening is the result of hindering the motion of dislocations by microstructural obstacles such as other dislocations, grain boundaries, sub-grain boundaries, solute atoms, particles of a second phase, and surface films. In cold working conditions, dislocations themselves are the most important obstacle to their own motion, leading to an increase in the material's flow stress. The hindrance of dislocation motion by microstructural obstacles also leads to an increase in the density of dislocations, further increasing the material's flow stress. This process can continue until the material reaches a critical dislocation density, at which point the material becomes too hard to deform further, leading to the phenomenon known as "work hardening saturation." When two dislocations approach each other, elastic interactions occur, leading to a barrier that must be overcome for the dislocations to glide in a slip system. This barrier is called the "Peach–Koehler force," which depends on the distance between the dislocations and their relative orientation. The barrier increases with an increase in dislocation density (ρ), which results in an increase in the material's flow stress (τ_c).

$$\tau_c = \tau_0 + \alpha G_0 \sqrt{\rho} \qquad (1.23)$$

Where α and G_0 are material constraints and τ_0 is intrinsic friction stress. For low-staking fault materials and at low strain, such as Cu-Al alloys, Cu-Zn alloys and austenitic stainless steels, dislocation dissociates in a confined slip planes so that localize cross-slip out of them is difficult leading to planner dislocation structure. At higher strain, these materials undergo extensive micro-twinning and shear banding. Plastic deformation can be illustrated in four states:

A) Stage-I: The stage-I of plastic deformation occurs at low strain levels and is primarily observed in single crystals oriented for single slip. During this stage, dislocations are confined to their slip plane and do not interact with each other. The crystal rotates by plastic deformation and tends to reorient towards a double-slip orientation, which leads to a lower energy state.

  In single crystals, there is no elastic interaction between dislocations, which leads to a low rate of work hardening. This is because the dislocations are not hindered by other dislocations or obstacles, and they can move relatively freely. However, in polycrystalline materials, Stage-I cannot be observed since there are multiple slip planes and dislocations are not confined to a single plane. The presence of multiple slip planes leads to dislocation interactions and an increase in the work hardening rate. In Stage-I, the plastic deformation is reversible, which means that the material can return to its original shape after the stress is removed. The dislocations are not able to multiply during this stage, which limits the amount of plastic deformation that can occur.

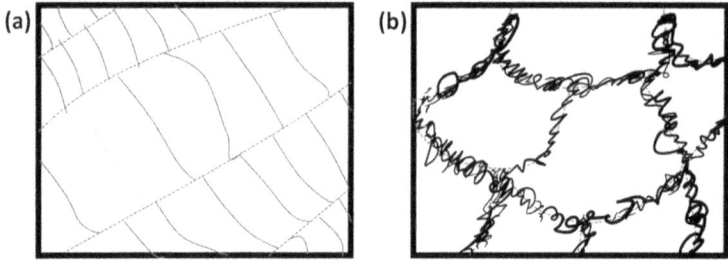

Figure 1.4 Schematic display of (a) dislocation pile-up or Taylor network and (b) dislocation tangles.

B) Stage-II ($0.05 < \varepsilon < 0.2$): A comparatively stronger dislocation interaction occurs in stage-II. Further, the movement of the first dislocation interaction is restricted by grain boundaries. Dislocation from different slip systems interact during dislocation multiplication leading to a substantial increase in dislocation density. Work hardening rates are high, being in the range of $30–40 \times 10^{-4}$ G. Low staking fault energy (SFE) materials exhibit a Taylor network (Figure 1.4(a)) whereas dislocation tangles are observed in high FFE materials, leading to the development of a cellular pattern (Figure 1.4(b)). The stage-II work hardening is independent of temperature.

C) Stage-III ($0.2 < \varepsilon < 1$): in this stage of work hardening, flow curve become parabolic i.e. the work hardening rate continuously decreases and the rate of work hardening is one order lower in magnitude than state-II. The primary mechanism is the balancing of dislocation multiplication (the continuation of Stage-II) and the local annihilation of dislocation. The annihilation of dislocation takes place by a dynamic recovery process due to a dislocation climb followed by the disappearance of neighbor dislocations having an opposite sign. The dislocation climb is a temperature-dependent phenomenon and, hence, the hardening rate in State-III varies with temperature. Microstructure evolution during this stage contains a well-defined cell substructure composed of dislocation cell walls. The microstructure contains complex tangles of dislocation that collapse into thinner and neater boundaries (Figure 1.4(b)). The grain interiors have low dislocation density. The grain dimensions ($0.1–1$ μm) decrease with the increase in plastic deformation. Further, the misorientation angle between the nearest cells increase in the range of $1–4^{0}$ with an increase in plastic deformation.

D) Stage-IV ($\varepsilon > 1$): with the increase in strain, the cell structure developed in State-III breaks down and causes the formation of microbands that are oriented in the direction of deformation. For example, grains are oriented in the direction of rolling during the metal forming operation.

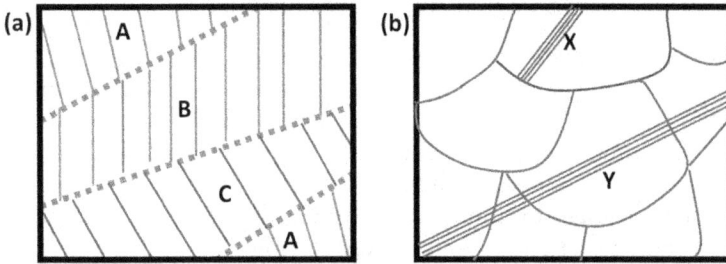

Figure 1.5 Schematic illustration of (a) deformation bands with orientation of crystallographic planes, and (b) shear bands showing different types of bands.

With an increase in strain, the oriented cell structure becomes a lamellar structure containing misoriented microbands in the direction of the rolling process. This facilitates an increase in grains and grain boundaries, leading to the production of a fibrous structure. Due to increased dislocation density and interaction with grain boundaries, the work hardening rate is higher (in the range of $1 \times 10^{-4}G$) than observed with large strains. At high strain, plastic deformation tends to localize as a deformation band and a shear band (Figure 1.5). The deformation band (Figure 1.5(a)) is the region within a grain, which deforms homogeneous structures, but with slip characteristics that are different from the neighboring regions. The stress state varies with the orientation of the deformation band due to the interaction of dislocations with grain boundary and secondary obstacles. These bands tend to align along the direction of straining. At higher strain, they form lamellar boundaries that facilitate the formation of fibrous structures. In contrast to shear banding, the deformation band is the manifestation of plastic deformation which can be explained by microstructural changes, but it does not constitute a form of damage to the material. Flow localization in form of a shear band develops when resolved shear stress in the fine lamellae is not sufficient for slip. Flat sheet-like structures composed of elongated crystallites, from 0.02 to 0.1 μm in width, are usually inclined at some angles to the rolling plane [3]. Shear bands are the regions of strong localized shears that often extend across the grains. The shear band initially develops at the grain scale and cross the grain boundaries with an increase in strain, eventually becoming macroscopic bands as shown in Figure 1.5(b). They evolve as diffused bands when the hardening rate decreases and the material is subjected to high straining. They become concentrated and become localized bands at high strain leading to failure. Materials having inhomogeneity due to their chemical composition (alloy precipitation, elemental segregation etc.) and crystallographic structure (twinning probability, low staking fault etc.) are prone to shear banding.

1.6.2 Work hardening at high temperature ($T > 0.4T_m$)

At higher homologous temperatures ($T > 0.4T_m$), thermally activated processes and strain rate can influence the flow behavior of materials during SSAM. At these higher temperatures, atomic diffusion becomes significant, and the dynamic recovery mechanisms, involving dislocation climb and cross-slip, become important factors in work hardening. During SSAM at high temperatures, dislocations can overcome the obstacles that hinder their motion, leading to a decrease in the dislocation density and a reduction in work hardening. This process is known as dynamic recovery, and involves the annihilation of dislocations and the rearrangement of the remaining dislocations. Atomic diffusion and corresponding dynamic recovery mechanism by dislocation climb and cross-slip reduces the flow stress.

$$\sigma = Z = \dot{\varepsilon} e^{\left(\frac{Q}{RT}\right)} \tag{1.24}$$

Resistant to atomic diffusion and dislocation motion increases with alloying and hence, the value of Q increases with alloying. Therefore, in alloys, dynamic recovery kinetics slow down even at elevated temperatures. Corresponding flow stress increases significantly, a phenomenon which is associated with an increase in dislocation density. The dislocation density easily attains a critical value for dynamic recrystallization during the hot deformation process. In the low Z regime, the saturation flow stress follows a power law:

$$\sigma_s = AZ^m \tag{1.25}$$

At high temperatures, dislocations can be rapidly annihilated. Solute atoms diffuse with dislocation (solute drag) and thereby control the rate of dynamic recovery and the formation of dislocation sub-structure [18]. Hence, dynamic recovery limits the dislocation pile-up and flow stress is controlled by the sub-grain size (d_{sg}). The microstructural features, including sub-grain size, depends on temperature, strain rate and solute content. During deformation, the sub-grains are continuously annihilated and replaced by new sub-grains of different size via the dynamic polygonization of free dislocations [19]. This maintains equiaxed sub-grain size and flow stress value over very large strains.

1.7 SOFTENING MECHANISM DURING SEVERE DEFORMATION

During severe deformation, softening can occur as the result of several mechanisms such as dynamic recovery (DR), dynamic recrystallization

(DRX) and grain growth. These mechanisms are related to the reduction of dislocation density and the formation of new grains, leading to a reduction in the flow stress of the material. The specific mechanism that dominates depends on the deformation conditions, such as temperature, strain rate, and strain. Solid-state deformation of material results in temperature increases and plastic deformation. Typically, 90% of processing energy is dissipated as heat, leading to an increase in the peak temperature of the process. The remaining energy (10%) is stored in the material, which results in microstructure change (deformed structure) and an increase in dislocation density. The deformation structure developed during the solid-state deformation process (the thermomechanical process) is subjected to annealing by thermally activated processes when there is no phase transformation. The deformation structure usually consists of an abundance of dislocation and having high stored energy due to strain hardening. Annealing of the deformed structure causes a reduction in the amount of stored energy. Surface energy/grain boundary energy is another driving force for microstructure change during the annealing process. Two important annealing mechanisms that reduce the stored energy are recovery and recrystallization. The activation of these mechanisms depends on material chemistry, staking fault energy, processing history and the level of stored energy (dislocation density). The recovery and recrystallization soften the plastically deformed materials, which leads to a reduction in strength and hardness (Figure 1.6) and an improvement in ductility.

During warm temperature ($T \leq 0.4T_m$; T_m = melting temperature of material) deformation, the kinetics for recovery is faster than recrystallization, leading to a slow reduction in hardness. At higher temperatures ($T > 0.7T_m$), recrystallization phenomena dominate the annealing process due to higher atomic diffusion. Recrystallization results in the formation of new strain free grain and, hence, the rate of hardness decrease is rapid. At intermediate

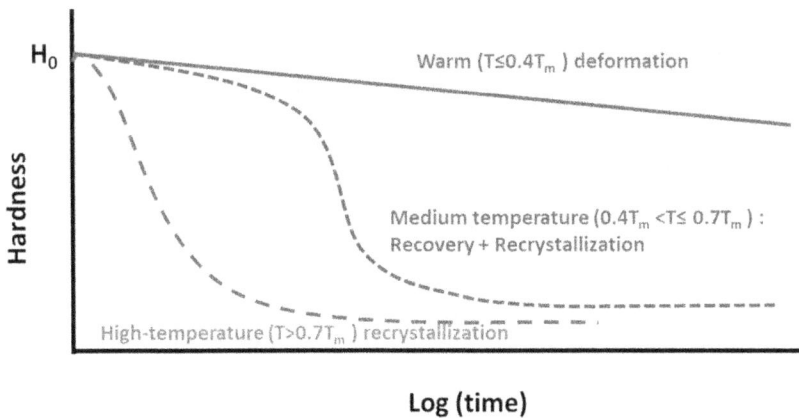

Figure 1.6 Schematic depiction of annealing kinetics at three temperature ranges.

temperatures $(0.4T_m < T \leq 0.7T_m)$, both recovery and recrystallization dominate the annealing process, causing a moderate decrease in hardness as shown in Figure 1.6.

During thermomechanical processing or solid-state deformation at higher homologous temperature, processed material experiences both strain-based deformation and high homologous temperature. For example, friction-based additive manufacturing experiences temperatures close as $0.8T_m$ and a strain rate of $10s^{-1}$ [20–22]. Processing could also occur at low homologous temperatures. For example, accumulative roll bonding and metal forming-based processing and cold welding can all be carried out at low temperatures. The softening mechanisms are usually demarcated into recovery, recrystallization and grain growth [23]. Each of these mechanisms needs a detailed discussion as follows.

1.7.1 Recovery

Recovery and dynamic recovery (DR) is a process that can occur during solid-state additive manufacturing (SSAM) processes, such as hot pressing or roll bonding. It is a microstructural change that occurs when the material is deformed at high strain rates and high temperatures. During DR, dislocations are pinned in the material, and the stored energy in the dislocations is released, leading to a reduction in the flow stress and an increase in the ductility of the material. The onset of dynamic recovery can be affected by several factors, such as the initial microstructure of the material, the deformation rate, and the temperature. Annealing by recovery occurs when there is a reduction of point and line defect (dislocation) at appropriate temperature. Any analysis regarding point defect during solid-state deformation can be ignored as the processing temperature is generally more than during a thermomechanical process $(T > 0.3T_m)$. It is important to note here that, in most materials, there is no boundary of demarcation between recovery and recrystallization. The so-called boundaries vary with material property and processing condition (stress, strain and temperature). Recovery is considered the most preferred mechanism of microstructure evolution at low homologous temperatures, whereas recrystallization is dominant at high temperatures. Both of the mechanisms are active in temperature regimes, as discussed in Figure 1.6. However, the relative kinetics of recovery and recrystallization varies with temperature and residual stress level. Recovery is a most active mechanism for the materials with higher staking fault energy (SFE), such as aluminum. High SFE favors cross-slip and therefore the local annihilation of dislocations, having an opposite nature in the same slip plane. There are four easy micro-mechanisms for recovery as shown in Figure 1.7. These include two types of dislocation (Figure 1.7(a)), namely i) positive dislocation and ii) negative dislocation. These dislocations are distributed in the microstructure. During annealing, dislocations become rearranged due to dislocation climb and cross-slip. When two dislocations having opposite signs and residing in a single slip place come together, their atomic defects

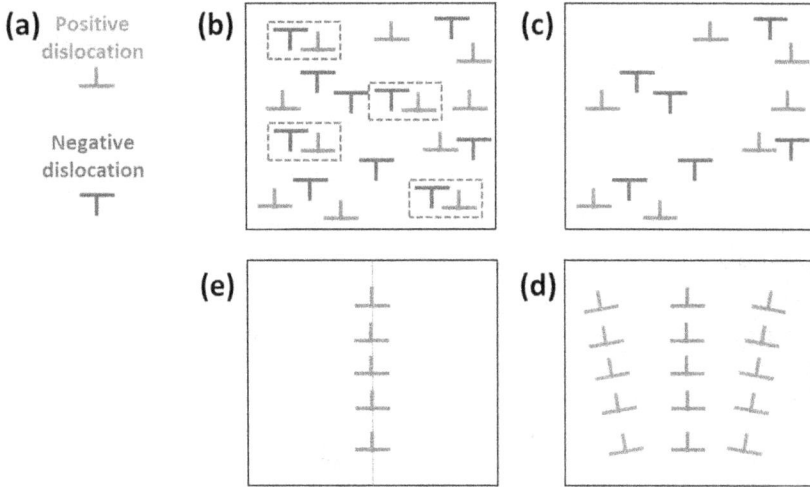

Figure 1.7 Schematic recovery mechanism showing (a) type of dislocations, (b) distribution of dislocation in the deformed microstructure and their rearrangement, (c) annihilation of dislocation having opposite sign, (d) polygonization and (e) coalescence of sub-boundaries and their growth.

rearrange (Figure 1.7(b)), leading to the annihilation of dislocation and reduction in dislocation density (Figure 1.7(c)). The remaining dislocation reorganized themselves into dislocation walls or sub-boundaries, a process which is called polygonization (Figure 1.7(d)). The rearrangement of dislocation in high SFE materials leads to the formation of dislocation cell walls and then sub-grain boundaries, which is having low energy configuration. In the final stage of recovery, the sub-grain grows during continued annealing, which leads to a reduction in internal energy. Experimentally observed kinetics of sub-grain growth can be expressed as follows:

$$d_s^n - d_o^n = kt \tag{1.26}$$

Where d_s and d_o are the final and initial sub-grain sizes. The k and n are constants that are dependent on temperature and varies according to the purity of material. The final sub-grain size also influences the flow and strengthening behavior of material.

$$\sigma - \sigma_0 = k_1 d_s^{-c} \tag{1.27}$$

Where σ and d_s are flow stress and final sub-grain size, respectively, and σ_0, k_1 and c are constants. This is the classical Hall–Petch type equation with $c = -0.5$ and 1 for high energy and low energy dislocation sub-structure, respectively. There is no clear distinction among all the process and hence it is difficult to analyze. Solute atoms in the alloy diffuse in the dislocation core and reduce the energy of dislocation, which eventually retards the dislocation mobility and therefore the recovery kinetics.

During the solid-state deformation process, material is subjected to cold working and stain hardening, which is experimentally observed during the typical uniaxial stress-strain curve. This strengthening occurs because of dislocation movements and dislocation generation within the crystal structure of the material. The deformation increases the dislocation density, which causes the evolution of low-angle grain boundaries surrounding sub-grains. During thermomechanical deformation, such as solid-state additive manufacturing, a reduction in dislocation density counterbalanced the work hardening by dynamic recovery and/or dynamic recrystallization. By contrast, during the dynamic recovery of high SFE materials, a steady-state flow curve can be seen due to the formation of sub-grains by a number of micromechanisms with the increase in flow stress and true strain, as seen in Figure 1.8(b). There are different stages of dynamic recovery as follows:

i) Initial microstructure subject to elastic deformation
ii) With an increase in strain, there is an increase in the flow stress as dislocations interact and multiply.
iii) With a further increase in strain, dislocation multiplication takes place. The driving force of recovery increases, resulting in the formation of low-angle grain boundaries and sub-grains in the microstructure.
iv) At a certain critical strain, the rate of work hardening, and recovery, reaches a dynamic equilibrium. Hence, the dislocation density remains constant, and a steady-state flow stress is obtained. The flow stress saturates after an initial period of work hardening and the saturation value depends on temperature, strain rate and composition.
v) With a further increase in strain, the original grains become increasingly strained, but the sub-boundaries remain more or less equiaxed. The substructure is 'dynamic' and re-adapts continuously to the increasing strain.

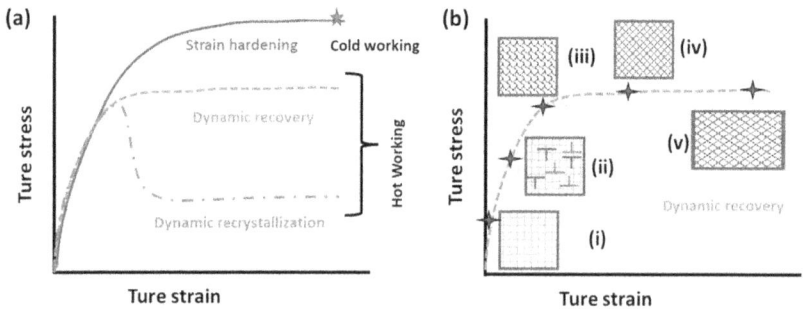

Figure 1.8 (a) A typical flow curve during cold and hot working conditions and (b) different stages of microstructure evolution during dynamic recovery.

1.7.2 Recrystallization

Recrystallization is a microstructural change that can occur during SSAM processes, such as friction-based solid-state additive manufacturing (FSAM). In addition to recrystallization, there is another related process, known as dynamic recrystallization (DRX), which can also occur during SSAM. Dynamic recrystallization is a process that occurs during high-temperature deformation when new grains form within the material due to the interaction of dislocations. Unlike recrystallization, which occurs after deformation has stopped, DRX occurs while the material is being deformed. During this process, the new grains form by a mechanism called strain-induced nucleation, which involves the formation of new nuclei due to the accumulation of stored energy from the deformation. Dynamic recrystallization plays an important role in SSAM because it can lead to the formation of new, equiaxed grains that can improve the mechanical properties of the final part. The rate of DRX is affected by several factors, such as the deformation rate, the temperature, the strain rate, and the initial microstructure of the material. For example, higher deformation rates and higher temperatures can promote dynamic recrystallization, while a larger initial grain size can inhibit the process.

The FSAM process reveals significant grain refinements due to dynamic recrystallization. The imposed high plastic strain overcomes the required critical strain for dynamic recrystallization, resulting in fine, equiaxed microstructural features consistent with wrought, rather than cast, materials [24]. The sub-grain formed during dynamic recovery grows by accumulation of free dislocation with further deformation. Then the sub-grain boundary is gradually converted to angle boundary (LAB) and high angle boundary (HAB) on the process of formation of a new strain-free and fine grain. Extended thermal exposure enlarges the grain due to grain growth mechanisms, which results in a decrease in strength and an increase in ductility. In the case of the multi-layer deposition of aluminum alloy, there is a refinement of equiaxed grain morphology in the deposited material. In addition, there is additional grain refinement at the interface between the deposited material and the substrate [25]. The grain size in the first layer is generally larger than that in the other layers due to more thermal recycling in the first layer. Recrystallization due to thermal exposure is due to static recrystallization. Since the temperature is low in solid-state deposition, kinetics for recrystallization due to temperature is very slow. However, the presence of dislocation density increases the kinetics because of higher atomic diffusion through dislocation and other defects present in the microstructure. Figure 1.8(a) shows that flow stress in recrystallization is very low compared to that in other mechanisms due to the formation of strain-free grain, which make the material ductile and soft. Figure 1.9 illustrates the microstructure changes during dynamic recrystallization. During solid-state deformation in additive manufacturing, material is subjected to thermomechanical processing on several

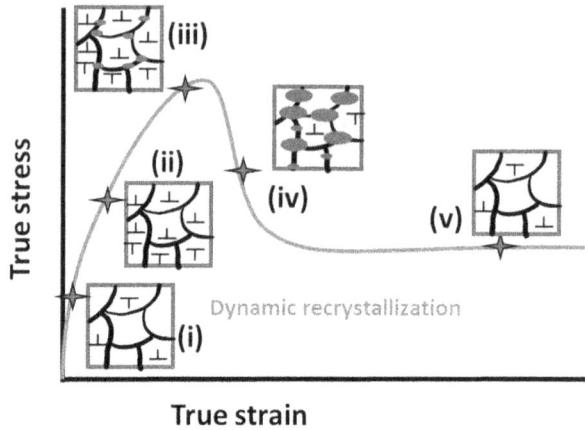

Figure 1.9 Schematic illustration on different stages of dynamic recrystallization with change of microstructure.

occasions depending on the type of additive manufacturing. For example, in FSAM, material at the boundary is re-stirred and is subjected to thermal exposure for the next layer of deposition. The reprocessing eliminates the possibility of defect formation at the interface between two subsequent layers. Hence, material that is dynamically recrystallized in the previous step of processing, is again dynamically recrystallized. It is important to note that, in the subsequent step of processing, the initial material which is going to be re-processed becomes soft due to dynamic recrystallization. Therefore, it is important to understand the stages of dynamic recrystallization as it helps in understanding the processing and predicting the mechanical properties.

As illustrated, there are five stages in the dynamic recrystallization of material at a defined critical strain:

i) During dynamic recrystallization, the grains, which had already recrystallized in a previous stage, are once again deformed. This initial grain always contains some fraction of dislocation.

ii) With an increase in deformation, dislocations interact and multiply, leading to an increase in dislocation density.

iii) At a critical strain, dynamically recrystallized grains appear at the old grain boundaries. New grains are subsequently nucleated at the boundaries of the growing grains – resulting in the so-called 'necklace structure'.

iv) With further deformation, more and more potential nuclei are activated at the prior grain boundaries. The evolution of strain-free recrystallized grain results in a reduction in the flow curve.

v) The new recrystallized grains replace the prior grain boundaries and appear throughout the microstructure. However, the dynamic process continues. Typically, equilibrium is reached between the hardening due to dislocation accumulation and the softening due to dynamic recrystallization. At this stage, the flow curve reaches a plateau and the microstructure consists of a dynamic mixture of grains with various dislocation densities. Unlike static recrystallization, the mean size of the dynamically recrystallized grains does not change as recrystallization proceeds.

1.7.3 Grain growth

Grain growth during solid-state deformation can be influenced by the same factors as those affecting recrystallization and dynamic recrystallization. However, grain growth occurs during or after the deformation process, and is usually slower than recrystallization or dynamic recrystallization. Grain growth can also occur during SSAM processes, and it can be affected by the same factors as those in solid-state deformation. For example, the temperature and cooling rate during the SSAM process can affect the rate and amount of grain growth in the final part. The use of proper process parameters can help to control the grain growth and improve the mechanical properties of the final part. Additionally, the presence of certain elements or impurities can also affect grain growth. For example, the addition of certain elements, such as aluminum or silicon, can retard grain growth, while the presence of impurities such as sulfur or oxygen can accelerate it. Furthermore, the type of deformation process can also impact the rate and extent of grain growth. For example, in hot deformation processes, the elevated temperature can promote grain growth, while in cold deformation processes, the deformation-induced strain can inhibit grain growth. Control of grain growth is important in many applications of solid-state deformation, particularly in the manufacturing of materials with specific mechanical properties. To minimize grain growth, various techniques can be employed such as controlling the deformation temperature, using alloying elements to inhibit grain growth, and employing heat treatments to promote recrystallization and refine grain size. Grain growth is the post-recovery or post-recrystallization mechanism that is driven by surface energy or grain boundary energy. Magnitude of growth kinetics is approximately two orders of magnitude lower than that of recrystallization. The factors affecting the grain growth are initial grain size, secondary pinning particle, temperature, thermal cycle and crystallographic texture [26–28]. Grain growth mechanisms can be demarcated into two broad categories: i) normal and ii) abnormal grain growth.

1.7.3.1 Normal grain growth

It occurs in polycrystalline materials to reduce the internal energy by reducing the total area of grain boundary for a homogeneous microstructure

Normal grain growth

(b)

(a)
Initial microstructure

(c)

Abnormal grain growth

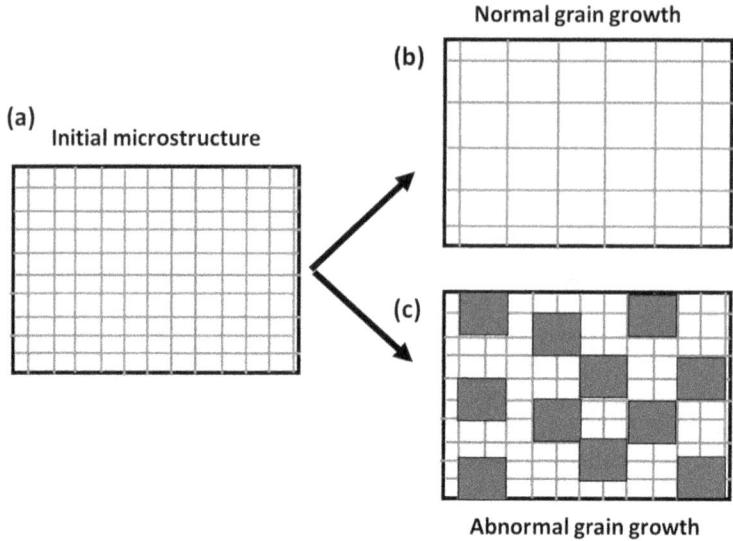

Figure 1.10 Schematic representation of microstructure after (a) recovery and/
or recrystallization, (b) normal grain growth and (c) abnormal grain
growth showing the difference in grain size and their distribution.

(Figure 1.10(a)). This appears analogous to the coarsening behaviors of
grains. Grain boundary is a defect in a crystal structure and, hence, it is associated with an energy. Therefore, there is a thermodynamic driving force to
reduce the total grain boundary area. With the increase in grain size by normal grain growth, there is a relative reduction in the actual grain boundaries
area in the microstructure (Figure 1.10(a, b)). In the general theory of grain
growth, normal grain growth can only occur in polycrystalline materials
with the full roughening of the grain boundaries. This can be approximated
by the following equation considering that the driving force is proportional
to the total amount of grain boundary energy.

$$d_f^2 - d_0^2 = kt \tag{1.28}$$

$$k = k_0 \exp\left(\frac{-Q}{RT}\right) \tag{1.29}$$

Where, d_f and d_0 are final and initial grain size, respectively; and k and t
are temperature-dependent constant and time of exposure, respectively.
Here, Q is activation energy for boundary mobility and self-diffusion. It is
important to note that these equations do not hold well for alloys but can
be applied in the case of pure materials.

1.7.3.2 Abnormal grain growth

Abnormal grain growth refers to the phenomena in which certain grains having energetically favorable grow rapidly and discontinuously in a refined microstructure, resulting in a distribution of grain size. The microstructure is dominated by a few large grains that are developed at the expense of their surrounding finer grains (Figure 1.10(c)). The driving force for the abnormal grain growth is an inhomogeneous microstructure with a variation in high energy sides due to high grain boundary energy, residual stress, locally high grain boundary mobility, favorable crystallographic texture or lower local second-phase particle density [26, 27, 29, 30]. This occurs primarily in alloys.

1.8 CASE STUDY: MICROSTRUCTURE EVOLUTION DURING FRICTION STIR ADDITIVE MANUFACTURING

The post-weld microstructure of friction stir processing (FSP) is characterized by analyzing the microstructure evolution of weld from the center to the base material (BM) of the weld. The deformation and the thermal gradient across the weld cause microstructural evolution as schematically shown in Figure 1.11(a). The FSP exhibits four conventional zones:

a) Stir zone (SZ): The stir zone (SZ) is the zone previously occupied by the tool pin. Typically, the diameters of the tool shoulder and tool pin determine the width of the nugget zone at the top and bottom of the weld. The thermal gradient and tool geometry decides the shape of the weld nugget (WN). In addition, heat input during friction stir welding (FSW) determines the spatial expansion of the WN. Deformation processes occurring in the material due to tool stirring dominate microstructural evolution. Furthermore, this zone experiences maximum temperature and plastic deformation. Therefore, this zone is generally characterized by a very fine-grained microstructure compared to that of the base material. The evolution of equiaxed grains is considered a signature of dynamic recrystallization involving high strain and strain rate deformation. It is important to note that within the WN there exists a strain and temperature gradient, and hence the grain size is not same at all the locations in the volume of the stir zone. The complex material flow, that is a characteristic of FSW, causes such variation in microstructural evolution. Complex tool shoulder and tool geometry increase the complexity of material flow, leading to additional flow patterns in the weld microstructure.

b) Thermo-mechanically affected zone (TMAZ): This zone surrounds the WN and has material that has undergone a low level of plastic deformation at lower strain rates. The plastically deformed material

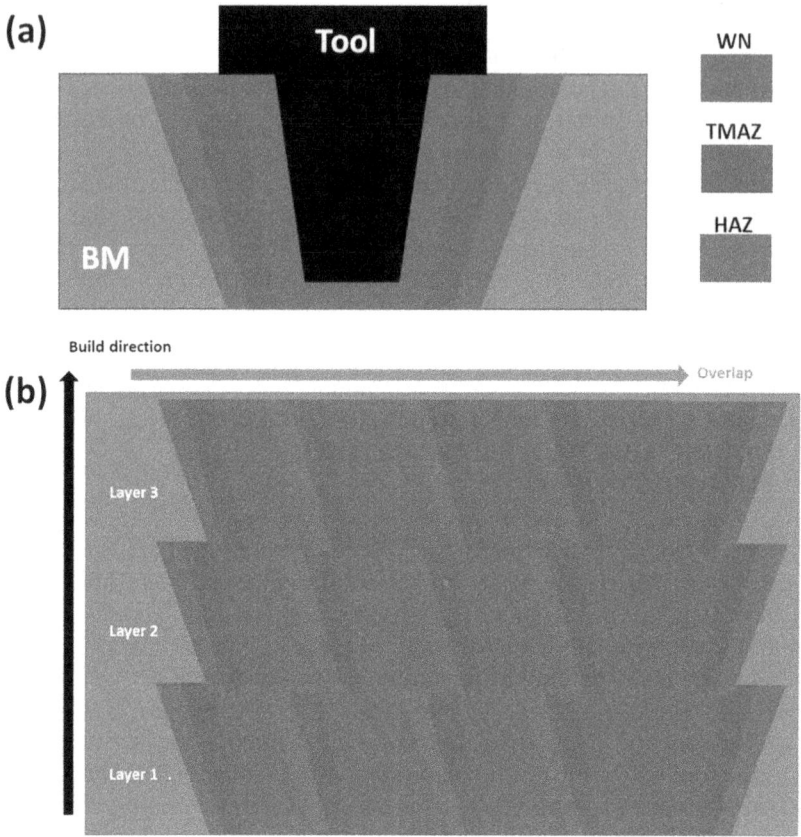

Figure 1.11 Schematic representation of (a) weld cross-section shows specific zones with respect to tool position and (b) macrostructure development during FSAM.

within this zone also experiences lower temperatures than that of WN. Since microstructural evolution depends on the deformation and temperature experienced by the material, this zone does not allow the evolution of a dynamically recrystallized fine grain structure. This zone is generally characterized by deformed and elongated grains containing high dislocation density. A distinct boundary exists between WN and TMAZ, depending on the temperature and the strain gradient, and causes a sharp boundary at the advancing size (AS) of the weld. A comparatively diffused boundary is observed at the retreating side (RS) of the weld due to lower temperature and strain gradient. The AS experiences a much steeper temperature and strain gradient than the RS of the weld as a new interface is formed at the AS due to the forward movement of the tool. Material that flows around the tool on the RS has to forge against this interface and form a weld.

c) The heat affected zone (HAZ): This zone exists between TMAZ and base material. The microstructure of this zone experiences only a thermal cycle, which leads to grain growth of base metal microstructure, especially in high thermal conductivity material such as aluminium. This zone is characterized by bigger grains when compared with the base material. Mahoney et al. [31] had reported that HAZ experienced temperature above 250°C for heat-treatable Al alloys. It is important to note that this zone suffers low-tensile properties due to grain growth and, hence, it is frequently subjected to tensile failure.

d) Base metal (BM): The extreme zone that does not experience both strain and temperature in FSW is regarded as the base metal (BM).

1.9 PREDICTION OF ADDITIVE STRUCTURE

Friction stir additive manufacturing (FSAM) is a solid-state additive manufacturing process that uses a rotating tool to stir and consolidate the material in a controlled way. This process can be used to create parts from a wide range of materials, including aluminum and titanium. During FSAM, the microstructure of the material can evolve in a number of ways, depending on the specific conditions of the process. For example, at low rotation speeds and low tool traverse speeds, the material experiences very little deformation and the microstructure remains relatively unchanged. However, at higher rotation speeds and tool traverse speeds, the material experiences more significant deformation, which can lead to the formation of new microstructures, such as equiaxed grains, sub-grains, and recrystallized regions. A deformation band, a region of high strain and high dislocation density, can develop as a result of the material's deformation during FSAM. The mechanical characteristics of the finished product, such as strength and ductility, can be significantly influenced by deformation bands. The initial microstructure of the material, the tool geometry, and the heat input can all have an impact on how the microstructure evolves during FSAM. In summary, the microstructure of the material can evolve during FSAM depending on the specific conditions of the process. The deformation of the material during FSAM can lead to the formation of new microstructures, such as equiaxed grains, sub-grains, and recrystallized regions and deformation bands, which can affect the mechanical properties of the final part. The microstructure evolution during FSAM can also be affected by the initial microstructure of the material, the tool geometry, and the heat input.

The microstructure evolution can be postulated during FSAM that can be considered as a derivative of FSP. Grain refinement in FSAM opens a new area of additive manufacturing to develop a structure with higher mechanical properties. In FSAM, the FSP process has been repeated along build direction and transverse direction as shown in Figure 1.11(b). Various microstructural regions in the FSAM are schematically shown in the figure. Processed zones are overlapped between two adjacent layers and, hence, transition zones are processed two times and experience higher thermal exposure. It produces very complex microstructure having many transition zones between two adjacent processed zones. Further, materials in the advancing and retreating side, which is soft and comparatively coarse due to grain growth, is subjected to further mechanical stirring and become refined. Hence, a homogeneous ultrafine microstructure with superior mechanical properties is obtained, as shown by Liu et al. [32]. Low temperature thermal cycle and rapid cooling during multi-pass FSP could suppress the undesirable grain coarsening in HAZ and obtain the similar grain sizes [33]. A similar microstructural feature has been detected for different materials, such as aluminum, magnesium, titanium and streel, with refined microstructure and marginal variation in mechanical properties along and across the deposition [11, 21].

1.10 ADVANTAGES AND DISADVANTAGES OF SOLID-STATE PROCESSING

Conventionally, fusion-based methods have been extensively used for metal additive manufacturing. Due to high temperatures in fusion additive manufacturing compared to that of solid-state additive manufacturing, fusion additive manufacturing samples usually suffers from porosity, residual stress, distortion, and evolution of intermetallic phases formation. Even though the solid-state method is unable to completely eradicate all problems associated with the fusion-based method, but it possesses a substantial advantage with regard to the absence of porosity and less distortion during solid-state welding along with the evolution of a lower fraction of intermetallics. It is important to note that welding by SSAM is costly compared to fusion-based methods, and these machines are seldom movable due to their requirement for high power, bulky space, and weight. Since solid-state additive manufacturing is the comparatively newly developed additive manufacturing method, it is essential to note both its advantages and its disadvantages.

A. The key benefits of solid-state additive manufacturing are:
   a) The solid-state method reduces problems related to solidification, such as distortion, and intermetallic formation. Since there is no melting involved, SSAM can produce parts with fewer pores and voids.
   b) Low temperatures and low distortion help to retain dimensional stability.
   c) SSAM can result in improved mechanical properties such as increased strength, ductility, and toughness due to grain refinement and enhanced microstructural characteristics.
   d) The process provides precise control over the manufacturing process, including the ability to produce complex geometries with high accuracy and repeatability.

e) Since the parts produced by SSAM are close to the final dimensions, they require minimal post-processing compared to other manufacturing methods.

f) As the process occurs in a controlled environment with minimal exposure to oxygen, there is a lower risk of the oxidation of the material.

g) No loss of process material during manufacturing and low temperature of the weld prevents the loss of alloying elements.

h) Composite and multilayer components can be fabricated, replacing joints comprising multiple parts and materials.

i) The process is environmentally friendly as it does not require shielding gas or additional consumables like flux and does not emit harmful gas or flux waste.

j) Hybrid structures reduce the weight of the component, decreasing fuel consumption in lightweight vehicles and aircraft.

B. Solid-state additive manufacturing also has a number of critical limitations such as

a) The loads experienced during SSAM are high and thus the machine used in this process is costlier and bulkier, restricting its commercial application and the fabrication of complex geometries.

b) The instrument is bulky and, hence, it cannot be used remotely. It requires considerable space and a high-power connection.

c) Since SSAM required special fixtures and tools, it cannot be used directly in certain complex designs.

d) A specially designed tool having high-temperature stability and wear resistance is essential for SSAM. The tool specification varies with material thickness and welding material. Therefore, a single tool cannot be used for different materials or thicknesses like fusion-based additive manufacturing.

e) There is a risk of abnormal grain growth in the stir zone during post-weld heat treatment, which could deteriorate the mechanical properties of the weld.

f) The process of SSAM is slower compared to fusion-based methods, which makes it less suitable for high-volume production.

g) SSAM may also face challenges with regards to the repeatability of the process and the quality of the final product, especially for complex geometries.

h) SSAM may have limitations in terms of the range of materials that can be used and may be unsuitable for all material combinations.

i) SSAM may also face challenges in terms of the quality of the weld, especially for dissimilar materials or for materials with different thermal expansion coefficients.

j) The residual stresses in the SSAM process can be significant, which may affect the mechanical properties of the final product.

k) SSAM may require post-weld heat treatment, which can increase the time and cost of the manufacturing process.

1.11 EFFECT OF SOLID-STATE ADDITIVE MANUFACTURING IN MECHANICAL PROPERTIES

Solid-state additive manufacturing can have a significant effect on the mechanical properties of a component. When opposed to earlier fusion-based processes, solid-state additive manufacturing can produce components with better properties due to the absence of porosity and intermetallic formation. The microstructure of the recrystallized weld nugget may enhance the mechanical properties of the component. The resulting grain structure is frequently finer and more regular compared to fusion-based methods, which can improve the weld's toughness, ductility, and strength. Moreover, solid-state additive manufacturing can produce parts with lower levels of residual stress and distortion, which can improve mechanical properties. Components may fit and align more accurately as a result of the reduced distortion, boosting their overall strength and longevity. The temperature and deformation rate can affect the emergence of novel microstructures, such as equiaxed grains, sub-grains, and recrystallized areas, which can alter the microstructure of SSAM sections. The material's future mechanical characteristics may be impacted by these microstructures. These microstructures can affect the following mechanical properties of the material:

a) Tensile strength: Because of the presence of recrystallized microstructures in the weld nugget and reduced porosity, tensile strength is typically higher in SSAM parts, resulting in improved material properties. However, as previously stated, the cooling rate during SSAM can also affect the microstructure and, as a result, tensile properties of the final component.

b) Ductility: The ductility of a material measures its ability to deform plastically without cracking. SSAM parts, in general, have lower ductility than conventionally manufactured parts. The cooling rate of the SSAM material can also affect the ductility of the final part. Rapid cooling can result in the formation of a fine-grained microstructure, which reduces ductility. Slow cooling, by contrast, can result in the formation of a coarse-grained microstructure, which improves ductility. In addition, the texture of the microstructure can affect ductility. Ductile materials, for example, have a more uniform and equiaxed grain structure than elongated and aligned grain structures. The orientation of the grains with respect to the direction of deformation may also influence ductility, with certain grain orientations promoting greater ductility. Furthermore, defects in SSAM parts, such as voids and cracks, can reduce ductility. Overall, SSAM part ductility can be improved by carefully controlling the processing parameters to achieve the desired microstructure while minimizing defect formation.

c) Hardness: The recrystallization of the microstructure during the SSAM process leads to smaller grain sizes and increased dislocation density, resulting in typically higher hardness of the SSAM parts. As a consequence, they exhibit greater strength and resistance to deformation and wear. However, the specific hardness of an SSAM part can vary depending on process parameters, material properties, and post-processing treatments.

d) Toughness: Toughness refers to a material's capacity to absorb energy before fracturing, which is commonly determined by evaluating the area under its stress-strain curve. Typically, SSAM parts exhibit lower toughness than parts fabricated by conventional manufacturing methods.

e) Fatigue and fracture: The toughness of a material can vary significantly, with metals having the highest values of fracture toughness, typically spanning four orders of magnitude. Tough materials such as metals are less susceptible to crack propagation, making them more resistant to failure under stress and resulting in a larger plastic flow zone in their stress-strain curve. Fatigue failures occur when metals undergo repetitive or fluctuating stress, leading to failure at lower stress than their tensile strength without plastic deformation. SSAM can reduce fracture susceptibility through a recrystallized microstructure, but optimizing the process to reduce interlayer defects is crucial. High strain rates and temperatures during SSAM can cause non-uniform microstructures and increase fatigue failure risk. The deformation rate and temperature affect new microstructure formation, impacting material toughness and fatigue resistance. The stress-strain diagram assesses fatigue resistance, strength, stiffness, and ductility under different loads.

f) Cooling rate: Slower cooling promotes atom diffusion and larger grain formation, resulting in a coarser microstructure with improved ductility, toughness, and creep resistance. Rapid cooling, on the other hand, restricts atomic diffusion, resulting in finer grains and increased strength and hardness. A finer microstructure, on the other hand, may result in lower ductility, toughness, and creep resistance. Controlling the cooling rate is therefore critical for achieving the desired microstructure and mechanical properties for specific applications.

In summary, the mechanical properties of SSAM parts depend on various factors, including the process conditions and the microstructure of the material. Compared to parts produced using traditional manufacturing methods, SSAM parts typically exhibit higher strength and hardness but lower ductility, toughness, and creep resistance. The final mechanical properties of the part can be influenced by the microstructure and cooling rate during the SSAM process. It is important to note that SSAM is a relatively new manufacturing process, and ongoing research is being conducted to improve the mechanical properties of SSAM parts and optimize the process for different materials and applications.

1.12 POTENTIAL APPLICATION OF SOLID-STATE ADDITIVE MANUFACTURING

The manufacturing industry is experiencing a technological revolution through the implementation of SSAM technology. SSAM allows for the on-demand manufacturing of parts, reducing supply chain length, storage requirements, delivery costs, and lead-time for critical replacement parts. The AM industry will continue to grow, with a projected sale of over AU$22.9 (US$15.8) billion worldwide across various sectors [34]. One of the key benefits of metal SSAM technology is its ability to improve manufacturing capabilities. Shorter lead-times, access to new materials, reduced material waste, and the ability to fabricate complex geometries and difficult-to-machine materials are all examples of this [34, 35]. SSAM has recently proven its effectiveness in a variety of industries, paving the way for a revolutionary shift in manufacturing processes. SSAM, which is capable of producing intricate and lightweight components, can benefit the aerospace and defense industries. For example, turbine blades for aircraft engines and gas turbine power plants can be made from SSAM, which has a high strength-to-weight ratio. SSAM can also manufacture high-temperature rocket engine components such as nozzles, injectors, and combustion chambers. Furthermore, SSAM can produce lightweight and robust satellite and spacecraft components.

The field of additive manufacturing is growing rapidly, and it is expected to produce around 80% of the overall production in the near future [36]. Although FSAM is still in its early stages, it has shown promise as a technology with excellent mechanical properties, microstructural advantages, structural efficiency, and eco-friendly processing [37]. These qualities make the FSAM process more adaptable in the future for engineering applications. It can be used in several fields of engineering sectors, including the development of 3D physical and functional products for automobile and aerospace, electronics, and industrial needs. Additive-based friction stirring has also proved its ability in the medical field, where it has been used to manufacture medical devices, surgical instruments, dental needs, knee joints, artificial body parts, and lab equipment. Although there is limited literature available on FSAM utilization in the medical field, it is estimated that it will cover at least 10–15% of medical needs in the near future. Based on past data [37], the current scenario, and future aspects, FSAM is expected to be a suitable alternative method of the additive manufacturing process for various applications. Its growth will continue to increase with research-based developments in the near future. These are just a few examples of the many potential applications of SSAM. As the technology continues to advance, it is likely that new applications for SSAM will be developed in the future.

1.13 CHALLENGES IN SOLID-STATE ADDITIVE MANUFACTURING

Solid-state additive manufacturing (SSAM) processes have become a popular method of producing complex components with a high level of accuracy, but the technology is not without its challenges. In particular, friction stir-based AM processes are relatively new and require innovation and optimization to overcome several challenges before they can be commercialized.

One of the main challenges associated with friction stir-based AM processes is the fabrication of intricate shapes. In contrast to fusion-based AM techniques, which enable the manufacturing of net-shaped complex geometries, friction stir-based AM processes require special fixtures for the fabrication of intricate shapes. The minimum feature size that can be printed using powder bed fusion is ~0.5 mm, whereas in the case of additive friction stir deposition, it is in the order of ~10 mm [38]. Improving in-plane resolution can be achieved by reducing the size of tooling and feed material. However, currently, the MELD machines (developed by MELD technology, Christiansburg, VA 24073, USA) for additive friction stir deposition have fixed feed rod size. Additive friction stir deposition poses significant challenges, including mechanical stability and tool wear [21]. The fabrication of high aspect ratio components can cause buckling, and thermal cycling during the process results in compliance gradient in the material, reducing the critical height–diameter ratio that can be printed without buckling. Post-process machining is necessary to compensate for the step-wise transition at the edge of each layer [21]. The issue of proper bonding near the edge also arises in side-by-side printing, which remains unexplored. The performance of friction stir-based additive manufacturing processes depends heavily on the tool's effectiveness in stirring the material. Tool health and lifetime are of utmost concern, especially for the processing of high-strength materials. The PCBN and W-Re tools are suitable for most materials used in friction stir-based AM, but expanding to high-temperature materials remains limited by the tool's strength, wear resistance, and fatigue resistance. A significant knowledge gap exists in quantitative relationships between process controls and resulting microstructure and properties, primarily due to a lack of process monitoring with feedback control capable of monitoring forces and temperature generated during deposition. A fully instrumented machine equipped with in-situ sensors can provide real-time dynamic feedback for process monitoring and control, as well as monitoring machine and tool health. However, these challenges are not unique to friction stir-based AM processes, and there are several other challenges that need to be addressed to secure the wider adoption and improved performance of SSAM. One of the key challenges is the lack of standardization in SSAM processes and materials, which can make it difficult to compare parts produced by different manufacturers or even within the same manufacturing facility. Material selection is also critical, and the materials used in SSAM must have a high melting point and low vapor pressure, as well as good thermal and mechanical properties. Some materials may require special handling or processing techniques, such as preheating or sintering, to achieve the desired properties in the final part.

To meet the required specifications, optimizing the process for each material and part geometry is critical, but it can be time-consuming and requires extensive experimentation and analysis. Metal printing presents a material anisotropy challenge, particularly with powders of varying particle sizes, shapes, and orientations. Depending on the printing direction, this can cause variations in the strength, ductility, and other properties of printed parts. Continuous research and development in the field of solid-state additive manufacturing is required to overcome these challenges.

REFERENCES

[1] M. Sriraman, M. Gonser, H.T. Fujii, S. Babu, M. Bloss, Thermal transients during processing of materials by very high power ultrasonic additive manufacturing, *Journal of Materials Processing Technology* 211(10) (2011) 1650–1657.

[2] G.J. Ram, C. Robinson, Y. Yang, B. Stucker, Use of ultrasonic consolidation for fabrication of multi-material structures, *Rapid Prototyping Journal* 13(4) (2007) 226–235.

[3] G.E. Dieter, D.J. Bacon, D. Bacon, *Mechanical Metallurgy*, McGraw-Hill (1988).

[4] A.K. Ghosh, A physically-based constitutive model for metal deformation, *Acta Metallurgica* 28(11) (1980) 1443–1465.

[5] C. Sellars, W.J. McTegart, On the mechanism of hot deformation, *Acta Metallurgica* 14 (1966) 1136–1138.

[6] D. Raabe, Deformation Processing, in: F. Bassani, G.L. Liedl, P. Wyder (Eds.), *Encyclopedia of Condensed Matter Physics*, Elsevier, Oxford, 2005, pp. 387–395.

[7] K. Hariharan, F. Barlat, Modified Kocks–Mecking–Estrin model to account nonlinear strain hardening, *Metallurgical and Materials Transactions A* 50(2) (2019) 513–517.

[8] V. Patel, W. Li, A. Vairis, V. Badheka, Recent development in friction stir processing as a solid-state grain refinement technique: Microstructural evolution and property enhancement, *Critical Reviews in Solid State and Materials Sciences* 44(5) (2019) 378–426.

[9] A. Kar, Y. Devinder, S. Suwas, S.V. Kailas, A study of deformation and microstructural evolution in friction stir welding of aluminum and titanium, *Trends in Welding Research* (S39-1) (2016) 728–731.

[10] A. Kar, S. Suwas, S.V. Kailas, Two-pass friction stir welding of aluminum alloy to titanium alloy: A simultaneous improvement in mechanical properties, *Materials Science and Engineering: A* 733 (2018) 199–210.

[11] S. Kumar, A. Kar, A review of solid-state additive manufacturing processes, *Transactions of the Indian National Academy of Engineering* 6(4) (2021) 955–973.

[12] A. Kar, S.V. Kailas, S. Suwas, Effect of Zinc interlayer in microstructure evolution and mechanical properties in dissimilar friction stir welding of Aluminum to Titanium, *Journal of Materials Engineering and Performance* 27 (2018) 6016–6026.

[13] A. Kar, S. Suwas, S.V. Kailas, *A Study of Microstructural Evolution in Friction Stir Welding of Aluminum and Titanium*, National Welding Seminer, Kolkata, India (2016).

[14] A. Kar, S. Suwas, S.V. Kailas, Multi-length scale characterization of microstructure evolution and its consequence on mechanical properties in dissimilar friction stir welding of Titanium to Aluminum, *Metallurgical and Materials Transactions A* 50 (2019) 5153–5173.

[15] Y.N. Wang, C.J. Lee, C.C. Huang, H.K. Lin, J.C. Huang, Influence from extrusion parameters on high strain rate and low temperature superplasticity of AZ series Mg-based alloys, *Materials Science Forum* 426–432 (2003) 2655–2660.

[16] C.I. Chang, C.J. Lee, J.C. Huang, Relationship between grain size and Zener–Holloman parameter during friction stir processing in AZ31 Mg alloys, *Scripta Materialia* 51(6) (2004) 509–514.

[17] Y.S. Sato, M. Urata, H. Kokawa, Parameters controlling microstructure and hardness during friction-stir welding of precipitation-hardenable aluminum alloy 6063, *Metallurgical and Materials Transactions A* 33(3) (2002) 625–635.

[18] A.A. Chularis, R.A. Rzaev, A.G. Valisheva, Formation and structure features of the weld joints made by friction stir welding, *Inorganic Materials: Applied Research* 10(3) (2019) 673–681.

[19] G.A. Malygin, Dislocation mechanism of dynamic polygonization of a crystal caused by its bending, *Physics of the Solid State* 44(7) (2002) 1305–1309.

[20] A. Arora, Z. Zhang, A. De, T. DebRoy, Strains and strain rates during friction stir welding, *Scripta Materialia* 61(9) (2009) 863–866.

[21] R.S. Mishra, R.S. Haridas, P. Agrawal, Friction stir-based additive manufacturing, *Science and Technology of Welding and Joining* 27(3) (2022) 141–165.

[22] A. Kar, D. Yadav, S. Suwas, S.V. Kailas, Role of plastic deformation mechanisms during the microstructural evolution and intermetallics formation in dissimilar friction stir weld, *Materials Characterization* 164 (2020) 110371.

[23] Chapter 5 Softening mechanisms, in: B. Verlinden, J. Driver, I. Samajdar, R.D. Doherty (Eds.), Pergamon Materials Series, Pergamon 2007, pp. 83–108.

[24] T.R. McNelley, S. Swaminathan, J.Q. Su, Recrystallization mechanisms during friction stir welding/processing of aluminum alloys, *Scripta Materialia* 58(5) (2008) 349–354.

[25] B.J. Phillips, D.Z. Avery, T. Liu, O.L. Rodriguez, C.J.T. Mason, J.B. Jordon, L.N. Brewer, P.G. Allison, Microstructure-deformation relationship of additive friction stir-deposition Al–Mg–Si, *Materialia* 7 (2019) 100387.

[26] A. Kar, S. Kailas, S. Suwas, Mechanism of variation in high-temperature grain stability of aluminum in dissimilar friction stir welds (2020) https://doi.org/10.1520/MPC20190011

[27] A. Kar, S. Suwas, S.V. Kailas, Microstructural modification and high-temperature grain stability of Aluminum in an Aluminum-Titanium friction stir weld with Zinc interlayer, *JOM* 71 (2018) 444–451.

[28] A. Sharma, V. Bandari, K. Ito, K. Kohama, M. Ramji, B.V. Himasekhar Sai, A new process for design and manufacture of tailor-made functionally graded composites through friction stir additive manufacturing, *Journal of Manufacturing Processes* 26 (2017) 122–130.

[29] D.A.H. Hanaor, W. Xu, M. Ferry, C.C. Sorrell, Abnormal grain growth of rutile TiO2 induced by ZrSiO4, *Journal of Crystal Growth* 359 (2012) 83–91.

[30] I. Vysotskiy, S. Malopheyev, S. Mironov, R. Kaibyshev, Pre-strain rolling as an effective tool for suppression of abnormal grain growth in friction-stir welded 6061 aluminum alloy, *Materials Science and Engineering: A* 733 (2018) 39–42.

[31] M.W. Mahoney, C.G. Rhodes, J.G. Flintoff, W.H. Bingel, R.A. Spurling, Properties of friction-stir-welded 7075 T651 aluminum, *Metallurgical and Materials Transactions A* 29(7) (1998) 1955–1964.

[32] M. Liu, B.B. Wang, X.H. An, P. Xue, F.C. Liu, L.H. Wu, D.R. Ni, B.L. Xiao, Z.Y. Ma, Friction stir additive manufacturing enabling scale-up of ultrafine-grained pure copper with superior mechanical properties, *Materials Science and Engineering: A* 857 (2022) 144088.

[33] Z.Y. Ma, S.R. Sharma, R.S. Mishra, Effect of multiple-pass friction stir processing on microstructure and tensile properties of a cast aluminum–silicon alloy, *Scripta Materialia* 54(9) (2006) 1623–1626.

[34] A. Vafadar, F. Guzzomi, A. Rassau, K. Hayward, Advances in metal additive manufacturing: A review of common processes, industrial applications, and current challenges, *Applied Sciences* 11(3) (2021) 1213.

[35] Y. Wang, Y. Lin, R.Y. Zhong, X. Xu, IoT-enabled cloud-based additive manufacturing platform to support rapid product development, *International Journal of Production Research* 57(12) (2019) 3975–3991.

[36] R. Liu, Z. Wang, T. Sparks, F. Liou, J. Newkirk, 13 - Aerospace applications of laser additive manufacturing, in: M. Brandt (Ed.), *Laser Additive Manufacturing*, Woodhead Publishing 2017, pp. 351–371.

[37] A. Kumar Srivastava, N. Kumar, A. Rai Dixit, Friction stir additive manufacturing – An innovative tool to enhance mechanical and microstructural properties, *Materials Science and Engineering: B* 263 (2021) 114832.

[38] H.Z. Yu, M.E. Jones, G.W. Brady, R.J. Griffiths, D. Garcia, H.A. Rauch, C.D. Cox, N. Hardwick, Non-beam-based metal additive manufacturing enabled by additive friction stir deposition, *Scripta Materialia* 153 (2018) 122–130.

Hybrid additive manufacturing using cold spraying and friction stir processing

Amlan Kar

South Dakota School of Mines & Technology, Rapid City, SD, USA

Sachin Kumar

NamTech Institute, Gandhinagar, India

Kuldeep Singh

Indian Institute of Science, Bengaluru, India

2.1 INTRODUCTION

Additive manufacturing (AM) is one of the rapidly developing technologies that have received substantial interest in various industries over the past few decades. It produces three-dimensional objects from digital models by depositing or hardening successive layers of materials [1]. Numerous almost net-shaped items are produced using AM techniques such as selective laser melt (SLM) and electronic beam additive melting (EBM), as well as 3D-directed energy deposition (DED). However, these materials still experience structural solidification problems and defects, and microporosity affects their plasticity and fatigue. Furthermore, these procedures are ineffective for metals with high reflectivity, such as magnesium (Mg) and aluminum (Al) [2].

The cold spraying (CS) process has been created to overcome the limitations of earlier AM techniques. It was initially developed in the Institute of Theoretical and Applied Mechanics of the Branch of Siberia published by the Novosibirsk branch of the Russian Academy of Sciences in the middle of the 1980s [3]. The spray comprises a gas pre-chamber cannon and a de Laval-style convergent-divergent propelling nozzle. A powder carrier gas from the gun's rear feeds the powder into the area upstream of the nozzle in an axial manner. By intake, the propellant is injected into the pre-chamber [1, 4]. To guarantee that the powder is appropriately incorporated into a gas combination, the operating pressure of the powdered form gas must be somewhat higher than that of the propulsion gas. The propellant gas particles accelerate the powder, reaching high speeds at low temperatures before colliding with the foundation. Ensuring feedstock powder particles collide with the substrate at speeds greater than a critical quality is the leading

DOI: 10.1201/9781032616025-2

technological objective of CS. Through the de Laval nozzle, pressure is applied at supersonic speeds to a pressurized and warmed propellant gas, such as nitrogen (N_2), atmosphere, or, in certain instances, helium (He). Through the use of high-pressure gas and relatively low temperatures to accelerate particles onto a substrate, CS avoids the solidification issues which frequently occur in conventional fusion-based AM methods. Research suggests that in the case of both Al and Mg AM, CS is a practical approach [5]. However, due to the porous, uneven microstructures and the lack of particle bonding, CS depositions have limited plasticity [6]. According to an earlier investigation, the porosity of CS deposition usually ranges around 5–12% [7], and the microstructure of CS 7075 deposit is non-uniform, with less than 3% elongation [8].

Friction stir processing (FSP), the second process considered in this chapter, is a highly localized extreme permanent deformation process that refines, homogenizes, and densifies material microstructures [8]. As a result, it can potentially increase strength while maintaining ductility. Previous research has illustrated that FSP may enormously transform CS coatings, boosting density [9], grain purification [10], and increasing resistance to corrosion [11]. According to Li et al. [12], the blend of cold spray and FSP, hereafter known as "CS+FSP," is a state-of-the-art hybrid AM technique for the production of lightweight metals. To date, however, there have been few investigations on preparing bulk light metals utilizing hybrid CS+FSP bulk composite additive manufacturing (hybrid CS+FSP). In recent studies, researchers explored the microstructure characteristics and mechanical behavior of an aluminum sample created using hybrid CS+FSP to show that this method can enhance the overall mechanical properties of lightweight metals. In this process, a high-pressure gas propelled Al particles onto a substrate, resulting in a dense, uniform coating. They then employed FSP to optimize the microstructural characteristics and enhance the mechanical behavior of the aluminum sample. The research revealed that the aluminum sample generated using hybrid CS+FSP had higher tensile strength, yield strength, and hardness than those prepared using standard AM procedures. Furthermore, the Al sample showed good ductility and toughness, which is critical for practical applications [13].

There are several benefits to employing hybrid CS+FSP technology over traditional AM techniques. It can swiftly produce near-net-shape products of various materials, making it appropriate for mass production. Second, it may prevent the usual solidification structure faults associated with standard AM processes, increasing the plasticity and fatigue performance of the products. Third, it allows for the preparation of highly reflective metals like Al and Mg which would otherwise be unsuitable for conventional AM techniques. Finally, hybrid CS+FSP technology can potentially improve the microstructure, density, and corrosion resistance of light metals, thereby boosting their overall mechanical qualities.

2.2 COLD SPRAY TECHNOLOGY IN ADDITIVE MANUFACTURING

Since the early 1980s, the additive manufacturing industry has used thermal spray processes, concentrating on manufacturing coatings for the hard tool sector. In this technique a spray gun is used to melt or semi-melt droplets of starting materials such as powdered form, wire, and rod, which are then fed into the gun's flame [14]. The type of energy source used in thermal sprayings, such as kinetic or thermal, can provide distinct properties to coatings or three-dimensional additive manufactured components. Thermal spray coatings are used in various scientific and technical sectors to improve the surface features of products for restoration and production, typically at a relatively low cost and with a lightweight and long-lasting investment. Conventional thermal spray methods, such as high-velocity oxygen-fuel (HVOF) and plasma spraying, produce coatings with little porosity and high adhesion [15]. Contrarily, conventional thermal spray techniques are usually impracticable for materials sensitive to temperatures or applications requiring coatings of a few millimeters thickness. This drawback was overcome with the invention of the gas dynamic cold spray coating method, frequently known as cold spraying (CS).

The discovery of cold spraying dates back to 1990 [16], when it was first recognized through a combination of experimental and theoretical analyses of supersonic dual-phase flow tests conducted in wind tunnels using a gas-solid particle mixture. It was initially created by a team of scientists from the Russian Academic of Sciences Siberian Branch's Institution of the Theoretical and Applied Mechanics (ITAM). These scientists proved that solid particle coatings can be generated at room temperature, avoiding heating to extreme degrees of temperature, as had previously been believed necessary. The method has been used successfully to repeatedly impact high-velocity metallic and metallic alloy particles [17]. The cold spray approach differs from conventional thermal spray methods since only the kinetic energy is used to produce coatings. The thermal spray method, by contrast, employs a mixture of heat and motion energies. Figure 2.1 depicts these processes and their difference in terms of the fundamental mechanisms involved.

2.2.1 Process and principle

Due to its capacity to produce homogenous coatings without heating the substrate, CS technology has grown in popularity in recent years. It achieves its results by combining a lesser processing temperature with a more significant high particle kinetic velocity. Unlike high-velocity oxygen fuel (HVOF) technology, cold spraying does not require particle melting. This allows it to be applied to temperature and oxygen-sensitive materials such as amorphous and nanostructured materials, in addition to copper (Cu), aluminum (Al), and titanium [18]. CS technology is classified into two types:

Thermal spray process = Thermal energy + Kinetic energy

Cold spray process = Kinetic energy

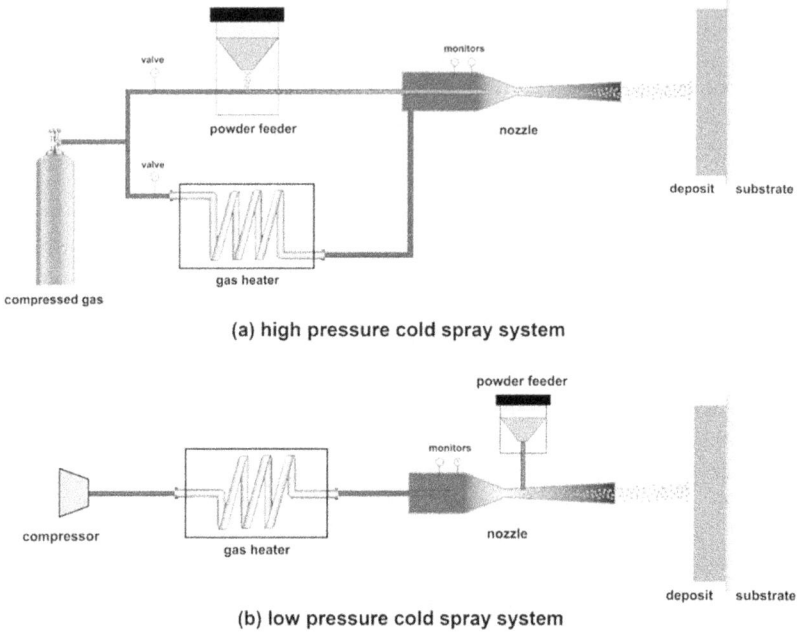

(a) high pressure cold spray system

(b) low pressure cold spray system

Figure 2.1 Schematic and mechanism of (a) a high-pressure CS system and (b) low-pressure cold spraying systems [1].

high-pressure CS and low-pressure CS [19, 20]. Propeller and carrier gas are the two gas streams of high-pressure CS. In this process the carrier gas is injected into a gas warmer as the propulsive gas is heated. After combining, the gas and powder streams are routed via a de-Laval nozzle to generate a supersonic stream. Low-pressure CS, contrastingly, uses a portable air compressor instead of a highly compressed gas. At the diverging nozzle portion, the gas pressure is so low that it allows powders dispensed by the powder feeder to be discharged at air pressure. Because of the use of lower pressures, CS systems are less costly and more adaptable. They have proved suitable for repairing the parts that are damaged. However, due to their lower particle velocity, they have limitations in material deposition and are mainly used for specialized materials like copper and aluminum. In the context of application sectors, cold spray coating and three-dimensional additive manufactured parts are in high demand in the maintenance and manufacturing industries. Given its an emphasis on high-pressure cermet-based CS coatings and components, the process mapping approach is commonly used in developing CS coatings and components.

2.2.2 Bonding mechanism

The impact of high-velocity particles and gas-depositing micron-sized solid particles on the substrate is utilized in the CS coating process. This process involves complicated material interactions, and the precise coating formation mechanism is unknown. However, a recent study dealt with the bonding mechanisms involved during CS [17, 21, 22]. Kinetic energy is converted into heat energy and mechanical deformation during the coating growth phase, which lasts 10^{-7} seconds [23, 24]. The bonding method, in which the substrate directly holds the particles, is generally attributed to mechanical interlocking. High velocities cause a significant temperature increase and bring about severe plastic deformation in the outermost layer of the particle when it comes into contact with the substrate. Localized melting of the particle results from this abrupt temperature change, and the molten material forges a solid bond with the substrate [25].

The theory of adiabatic shear instability states that thermal softening-induced contact among particles or separating the substrate from the particle may create localized adiabatic shear instabilities during impact. The thin oxide surface layers are disrupted or become unstable due to the significant localized distortion caused by this, particularly shear deformation. As a result, the particle comes into close conformal proximity to the substrate or earlier deposited layer, resulting in bonding. Due to localization, the flow stress instantly reduces to zero, which causes the insertion of an interface jet made of the same substance as the severely deformed particle. Finally, the jet eliminates the oxide coating of the particle and substrate, enabling direct contact with the metal surfaces. This interaction causes bonding, relatively elevated contact pressures, and interfacial softening [26].

Understanding the CS deposition mechanism requires knowledge of particle deformation. During the CS process of material deposition a substrate is subjected to many collisions of solid particles that are micron-sized. A feedstock powder is compressed into a deposit by the viscoelastic deformation of the contacting bodies and particle impact. Metallurgical bonding occurs on a significant amount of the particle-to-particle faces. The resulting particle impact and the high deformation are used in compaction and bonding procedures. Understanding particle deformation is therefore critical in understanding the CS deposition mechanism.

The heat produced by plastic deformation can cause strain localization and shear instability, especially at high strain rates. This phenomenon is referred to as either shear banding or failure in bulk materials [27]. The inherent instability of material causes localization and shear banding. Particle bonding in CS is expected to be reasonably comparable to this process. The overall strain is nearly constant up until breakage, but at the critical strain point the shear band's localized strain increases dramatically to high values [26, 28].

On the other hand, few substances may necessitate using other material models and analytical procedures. For instance, because strain softening occurs instead of work hardening, metallic glasses are unstable and prone to shear banding. However, some intermetallic compounds may resist instability due to anomalous (non-monotonic) thermal softening.

2.3 FRICTION STIR PROCESSING IN ADDITIVE MANUFACTURING

Heat is generated by friction using a high-speed machine, a shoulder, and a projecting pin in the welding process known as friction stir welding (FSW). This heat is necessary in order for the welded component to deform plastically. Friction at the point of contact produces a lot of heat when the tool spins and moves in relation to the substrate. This leads to the generation of a plastically deformed zone with low flow stress, high viscosity, and three-dimensional fluidity. The exclusive observation of these features is in the local third-body region [29]. When a tool that is either consumable or becomes so during operation is employed on the workpiece, traces of the third-body region might become discernible on the tool itself. As the tool generates the requisite frictional heat, it is a fundamental component of the process.

In Type 1 friction stir additive manufacturing (FSAM), rotational friction welding and friction deposition are used to generate the third-body area in the tool itself. The surface of the workpiece has been coated to allow the tool to spin around and move across it. The tool in the friction deposition method is a shoulder-connected hollow cylinder [23]. After being heated by the frictional resistance of the contact, the feed material is delivered to the substrate interface as a straight solid rod or powder. The surface of substrate, where the substance is placed, is traversed by the device. A three-dimensional building can be built by piling the layers on top of one another. With this method, the material is deposited after being plastically deformed from the tool and moving in a translative motion. The feedstock is given by rotational stirring. Additionally, functionally graded materials (FGMs) with changing compositions and properties in several component areas can be produced using powder feedstock in additive friction stirring (AFS). Considering that it may generate components with close-to-net shapes and configurable microstructures and properties, FSAM is a possible strategy for the aerospace, pharmaceutical, and energy sectors [24].

An inventive method of autogenous solid-state welding called friction stir additive manufacturing (FSAM) uses a tool that is not consumable. After creating a lap joint between two sheets, type 2 FSAM is created through the repeated application of FSW to link fresh sheets to an existing structure. The top plate has a customized, non-consumable rotational tool with precisely calibrated pin and shoulder specifications. In order to achieve a strong join, the pin depth is chosen to go through 20–30% of the lower plate, and other parameters are varied as needed. Like friction stir lap weld (FSLW), the FSAM process offers greater control over the duration of the warming and the re-sintering, allowing for better management of the material characteristic and desired mechanical features. It is a method that leads to parts with improved mechanical behavior, such as enhanced strength, fatigue resistance, and ductility. FSAM also provides numerous benefits over traditional production techniques, including the capacity to build complex shapes, minimize material waste, and integrate multiple materials [30, 31]. The process is used in a wide range of industries. For example, in the aerospace industry it is employed in the fabrication of large aircraft components, including wings and fuselage sections. Similarly, in the automotive sector it helps to create lightweight, strong, and rigid components such as suspension and engine elements. In the biomedical industry it can be employed to make implants with properties that promote bone formation and healing.

Prior study in hybrid additive manufacturing has demonstrated that FSP can effectively impact the properties of CS coatings, resulting in increased density [9], refinement of grains [10], and resistance to corrosion [11]. Li et al. [12] looked at the potential of "hybrid CS+FSP," an innovative hybrid technique for additive manufacturing for light metals that incorporates the strengths of CS and FSP. To date, however, there has been limited research on the production of bulk light metals adopting the hybrid CS+FSP technique. The hybrid CS+FSP design is shown in Figure 2.2(a), where cold spraying is used to produce an additive layer on the selected substrate. After that, the material is deposited, and multi-pass FSP is employed. The preceding two steps continued until the bulk sample is finished.

There are many benefits to employing hybrid CS+FSP technology rather than traditional AM methods since hybrid CS+FSP can create highly flexible materials and fatigue resistance while avoiding solidification structural faults. Second, it can be utilized to create metals with high reflectivities, like aluminum and magnesium, which are challenging to develop using the AM technology that is now available. Ultimately, hybrid CS+FSP coatings can have a high-density, homogeneous microstructure, and strong particle bonding. Finally, hybrid CS+FSP offers the advantages of both CS and FSP, yielding a novel and efficient method for additively producing light metals with exceptional mechanical properties. This hybrid method is particularly suitable for metal matrix composites where distribution of second-phase particles are of the utmost importance.

A new addition to the solid-state form additive manufacturing family is hybrid CS+FSP. Like AM, CS can be used to construct anything by layering freeform components [26, 33]. Unlike traditional AM techniques, hybrid CS+FSP generates relatively little heat on the spray materials (substrates).

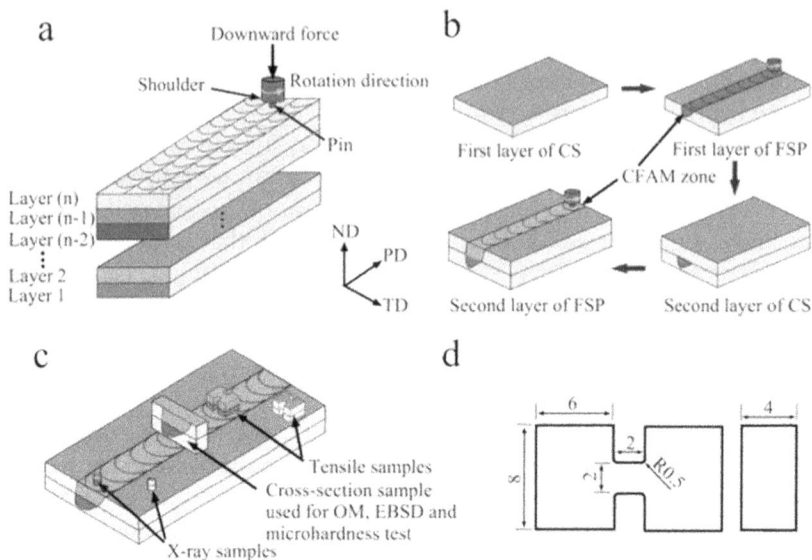

Figure 2.2 Schematic representation of (a) HYBRID CS+FSP, (b) experimental methodology, (c) location of sampling, and (d) dimensions of the tensile sample [32].

ND stands for "normal direction," and TD for "traverse direction." Unit: millimeters.

Further, hybrid CS+FSP may construct blocks and complicated freestanding structures faster than traditional AM techniques. The field of additive manufacturing has quickly expanded to generate practical and complex components for various industrial uses by fusing the additive aspect of hybrid CS+FSP with the subtractive results of conventional machining [26, 34]. To produce multi-metal deposits utilizing hybrid CS+FSP, a platform in motion was used in the 5-axis machining centers of Ibarmia and Hermle. Nonferrous metals have also been successfully joined together using this method [35]. This process can boost grain refinement, microstructure, corrosion, and the mechanical characteristics of the base material [36, 37]. The CS+FSP technology combines depositing material with CS and using FSP to improve microstructure and mechanical properties of materials while correcting flaws caused by the CS process [32]. The CS+FSP technology can expand its applications in the industry while overcoming the limitations of CS-AM.

2.4 MATERIALS

With an emphasis on lightweight metals like aluminum, magnesium, and titanium, the hybrid CS+FSP technology is a hybrid AM approach with a lot of potential. These metals are essential materials in many industries, including the aerospace, automotive, and healthcare sectors, due to their remarkable properties, which include relatively low density, excellent specific strength, and outstanding durability against corrosion.

2.4.1 Hybrid additive manufacturing of aluminum

2.4.1.1 Hybrid additive manufacturing of pure aluminum

Wang et al. [32] created a bulk pure Al sample (hybrid CS+FSP) via CS-FSP composite additive manufacturing. A cold spraying mechanism (LERMPS, UTBM, France) was utilized to deposit spherical pure aluminum powder particles with a typical size of 17.34 μm at 2.8 MPa nitrogen pressure and 350°C temperature. The deposition angle, moderate powder feed rate, nozzle stand-off length, and nozzle movement speed were all adjusted to 40 mm/s, 90°, and 2.5 rad/min, respectively. The sample was processed with FSP on an FSW-LM-AM16-2D machine using a tungsten carbide (WC-Co) tool with a 16-mm flattened shoulder and a 3-mm length pin with a tapered diameter at 5 mm at the shoulder to 2.5 mm at the pin end. The processing parameters were as follows: parallel to the CS direction is the processing direction (PD); 1500 rpm rotating speed; 23.5 mm/min traverse speed; tilt angle of 2°; and a shoulder plunging depth of 0.2 mm.

The variation of microstructure and microhardness is depicted in Figure 2.3(a). In the hybrid CS+FSP zone, the microhardness is 32 HV on average, which is 28% superior to the CS zone. The strengthening effects of the size of the grain, dislocations, grain position, and increased density cause this increase in microhardness. According to prior research, these components have an identical effect on tensile qualities. Figure 2.3(b) depicts the engineering stress-strain graph for the CS and hybrid CS+FSP samples. The results for the CS sample's yield strength (YS), ultimate tensile strength (UTS), and elongation (EL) are 24 MPa, 60 MPa, and 4.2%, respectively. The yield strength, ultimate strength, and elongation of the hybrid CS+FSP sample are 33 MPa, 87 MPa, and 60.3%, respectively, are 38%, 45%, and 1336% higher than the rest of the CS sample. The contribution of several strengthening methods to the strength of the CS and hybrid CS+FSP samples is shown in Figure 2.3(c). Because of its high porosity and weak interparticle bonds, the CS sample's experimentally determined yield strength is lower than its calculated values. The hybrid CS+FSP sample exhibits a higher rate of strain hardening and constant elongation of 19% when the engineering stress-strain graph of the CS and hybrid CS+FSP samples are contrasted in Figure 2.3(d). The CS sample falls short before the convergence of both plots because of its severe porosity. There is a brittle fracture on the fracture surface of the cold spray CS sample alongside the particle contact, as illustrated in Figure 2.3(e). On the flip side, ductile fracture of the hybrid CS+FSP sample exhibits a disproportionately high quantity of uniformly spaced

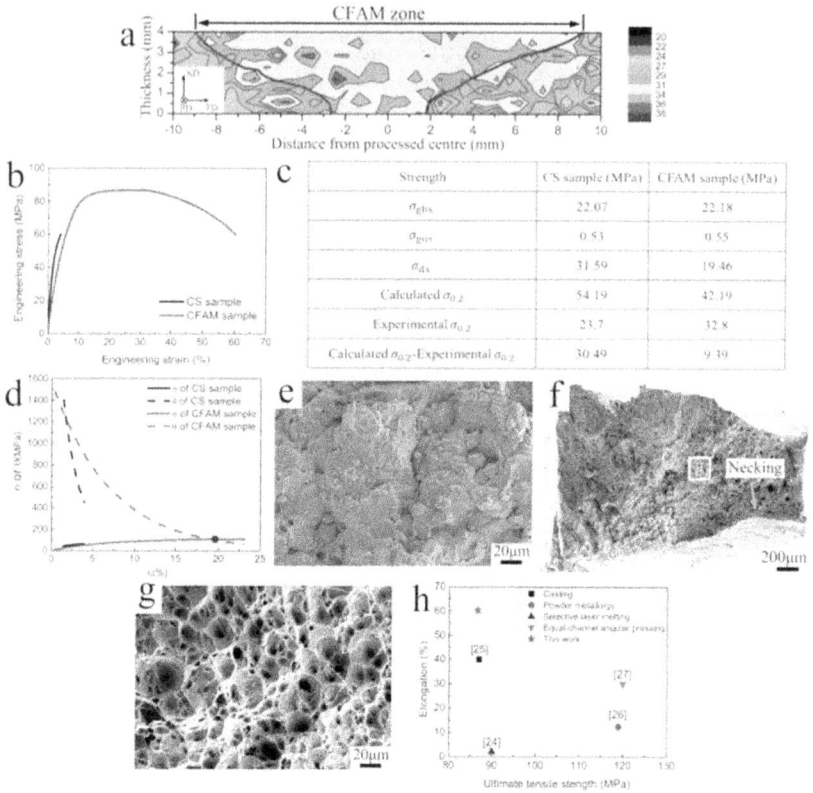

Figure 2.3 Depicts (a) the variation in microhardness; (b) engineering stress–strain graphs; (c) the role of different strengthening methods; (d) true stress–true strain graphs, working hardening rate–true strain curves, and the distribution of microhardness; (e) fracture surfaces of CS and HYBRID CS+FSP samples; (g) region of surface fracture g in (f); and (h) comparisons of the tensile properties of pure Al produced using different procedures [32, 38–40].

dimples (Figure 2.3(f) and (g)). These findings align with previous research, such as that of Read et al. [41]. The tensile properties of pure Al generated in various ways are shown in Figure 2.3(h). Because hybrid CS+FSP reduces cemented defects, the hybrid CS+FSP samples have a similar UTS but a noticeably higher EL compared with the as-cast and selectively laser-melting specimens (by 51% and 2915%, respectively). The strength of the hybrid CS+FSP sample is lowered by 382% when compared to samples produced by powder metallurgy with equal-channel angular pressing. However, a 101% rise in the EL reveals the high-density and strain-hardening characteristics of the hybrid CS+FSP pure Al. According to these results, combining CS and FSP is a potential method for generating pure bulk aluminum with improved mechanical behavior.

2.4.1.2 Hybrid manufacturing of AA7075 coating

Khodabakhshi et al. [42] examined friction stir processing (FSP) to change a CS AA7075 layer in order to improve structural uniformity, microstructure refinement, and hardness in a bi-material construction of coating on an AZ31B substrate Figure 2.4(a–d). The constituent phases of the cold-sprayed coat were blended and polished using high plastic deformation to achieve completely dense solid coating layers. Chemical bonds may occur, and intermetallic phases may exist at the contact. In various industrial applications, this method might enhance the performance of bimetallic structures.

Figure 2.5(a, b) displays the optical microstructures of the cold-sprayed covered layer on the magnesium-based substrate before and after etching. The as-sprayed sample had a final deposition thickness between 200 and 250 µm. The interface of AA7075 coating with the AZ31B substrate changed during the FSP modification from wavy to smoother, with some grain coarsening emerging under the shoulder. Compared to the initial substrate alloy, refining around the small revolving disc pin increased the coating layer's Vickers hardness by more than three times. The findings of the study indicate that during the cold-spraying process, high temperature and high-velocity impingement refined the microstructure of substrate close to the surface. The FSP caused the wavy interface with the AA7075 coating and the AZ31B substrate to transform more smoothly. The Vickers hardness of the coating layer was increased more than three times over the initial substrate alloy by refining around the tiny rotating pin. The study found that the high-temperature, high-velocity impingement cold-spray approach refined the microstructure of the substrate alloy near the surface. Following FSP modification, the coated layer's microstructural and mechanical characteristics were significantly enhanced. The microstructures shown in Figure 2.5(b), located toward the pin-side effect while retreating to the side, significantly improved the bonding mechanisms of the cold-sprayed metal surface and the substrate. Additionally, the tiny pin used in the FSP technique caused material mixing to create the textural profile illustrated in Figure 2.5(b).

2.4.1.3 Commercially available pure Al coating over pure Ti substrate

Friction stir spot processing (FSSP) was utilized by Ji et al. [43] for microscopic analysis, evolution microhardness measurements, and the adhesion strength of aluminum (Al) coatings cold-sprayed onto titanium (Ti) surfaces. The raw material's powdered form was a gas-atomized, readily accessible, pure aluminum alloy powder with a nominal particle size of 35 m. At the same time, the substrate was built of pure titanium plates (40 × 40 × 3 mm³). To remove any remaining dirt, the substrate was washed in an alcohol ultrasonic bath before being grit-blasted with alumina grits. The coating of pure Al was sprayed to the Ti substrate using cold-spraying equipment

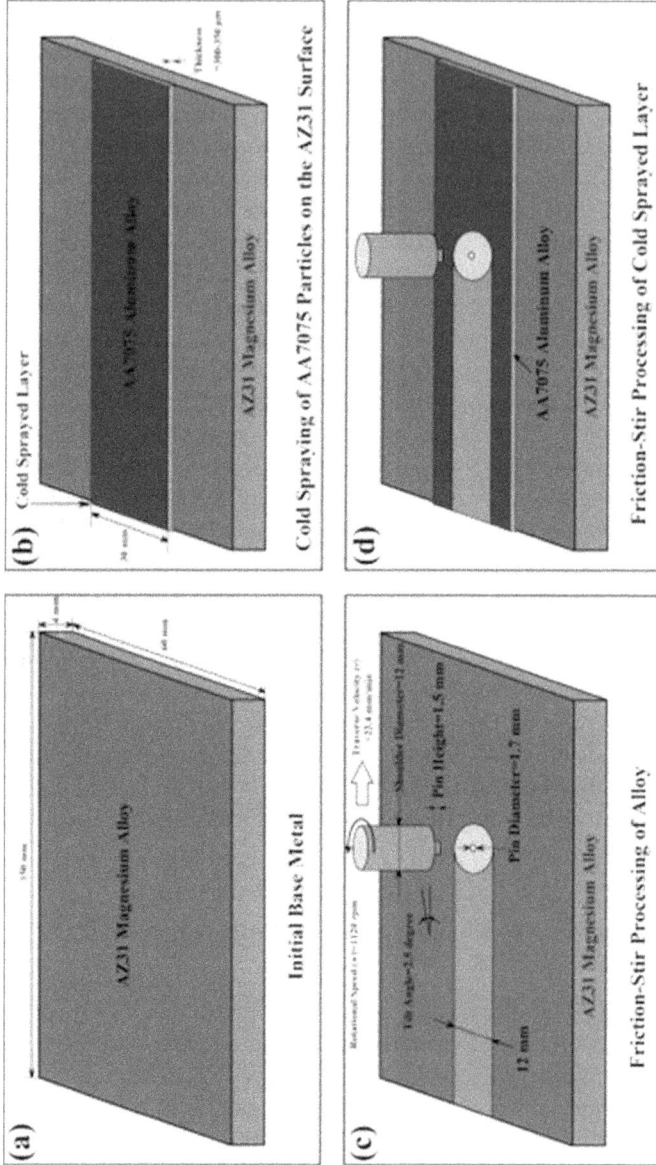

Figure 2.4 Schematic depictions of the (a) first substrate made of AZ31B alloy, (b) AA7075 layer being cold sprayed onto the AZ31B substrate, (c) original AZ31B alloy FSP, (d) AZ31B-AA7075 bi-materials structure FSP and final AZ31B-AA7075 alloy FSP [42].

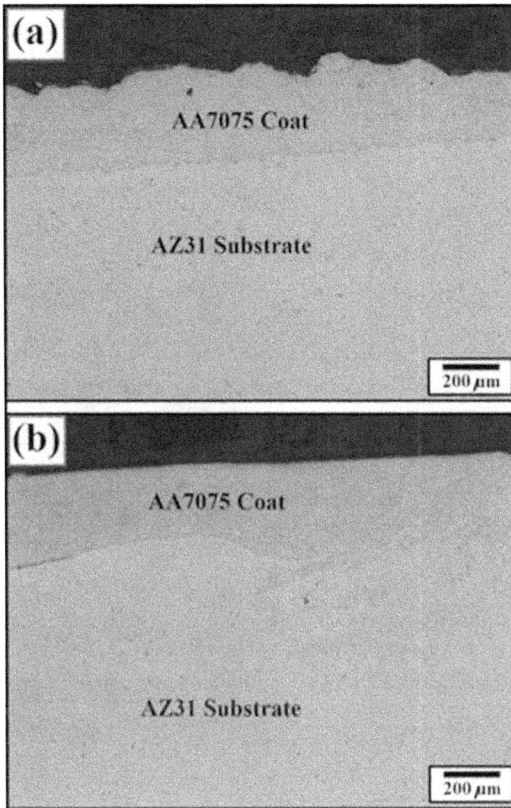

Figure 2.5 Cold-sprayed layer (a) before FSP and (b) after the FSP procedure [42].

(CS-2000, University of Xi'an Jiaotong, China) at 3 MPa and 230°C with nitrogen as the accelerating gas. The gun operated at 40 mm/s, and 30 g/min of powder was fed. Cold-sprayed Al coatings were thinned out before FSSP to a thickness of roughly 1.5 mm to smoothen the coating surface. This process was carried out with commercially available friction stir welding equipment (LQH-G15, Lianqiao Weihai precision tools Co. Ltd., China) at various rotation speeds (1800, 2100, 2400, and 3000 rpm) with plunge rates and depths and dwell periods of 1 mm/min, 0.5 mm, and 30 s, respectively. Six involute grooves were present on the H13 steel surface of the processing tool. Table 2.1 displays the outer layer and sectional morphologies of cold-sprayed aluminum coats before and after FSSP at various rotation speeds.

After FSSP at 1800 rpm, the cold-sprayed aluminum coating darkens, reducing inter-particle interfaces and real pores to produce a dense region. Due to insufficient heat input and material movement, the lush area experiences cavity flaws during FSSP, which may weaken the bonding strength and promote surface ripping. The cavity flaws in the 1/2 radius and dense area

Table 2.1 Aluminum coatings that are cold-sprayed surface characteristics and cross-sections at various rotation speeds before and after FSSP [43]

Tool rotation speed (rpm)	Surface morphologies	Cross-section images
As deposited		
1800		
2100		
2400		
2700		
3000		

center can be eliminated with enough heat input at a rotational speed increase of 2100 rpm. The hollow imperfections are still present around the border due to the reduced processing temperature. The additional heat is needed to provide an ideal cold-sprayed aluminum covering at 2400 rpm. However, when the rotational speed is increased to 2700 or 3000 rpm, cracks appear at the edge of the dense area due to the significant heat input created at high spinning rates during FSSP.

The cold-sprayed aluminum coating's particle morphology prior to and following FSSP at 2400 rpm are shown in Figure 2.6. In Figure 2.6(a), four zones are shown as a component of the treated coating: the SZ, the zone that is affected by heat (HAZ), the thermo-mechanical affected zone (TMAZ), and the unmodified zone (UAZ). Mechanical interlocking is clearly the dominant form of particle bonding, as evidenced by the unaffected zone (UAZ) sharing the same microstructure. The images in Figure 2.6(b, c) further substantiate this claim, demonstrating flattened particles, preserved pores, and direct particle-to-particle contacts. Frictional heat contacts the HAZ close to the UAZ in Figure 2.6(d), producing roughly spherical particles. The HAZ, however, still contains the original pores and inter-particle interactions. Figure 2.6(e) shows larger particles in the TMAZ, and the real

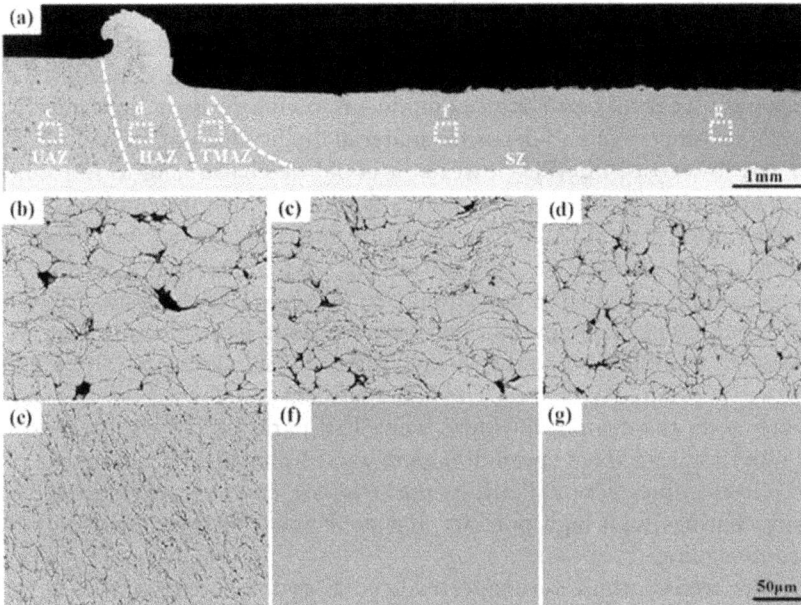

Figure 2.6 Al covering by cold-spraying particle morphologies prior to and following FSSP at 2400 rpm: (a) cold-sprayed aluminum coating; (b) as-
deposited Al covering macrostructure; (c) UAZ; (d) HAZ; (e) TMAZ;
(f) 1/2 radius of SZ; (g) center of SZ [43].

pores are gone. The continuous inter-particle interfaces progressively give
way to distributed micropores as temperature and plastic strain increase.
Figure 2.6(f–g) depicts the eradication of micropores in the SZ as a dark and
dense zone at the macroscopic level, indicating a shift toward the dominant
mechanism of metallurgical bonding between the particles. Such a bonding
after the hybrid CS+FSP is the primary reason for higher structural integrity
and mechanical properties.

2.4.2 Hybrid additive manufacturing of steel

Austenitic steels are utilized extensively in various industries due to their
remarkable durability against corrosion in severe environments and also
their powerful mechanical qualities at increased temperatures. For instance,
the fabrication of connecting pipes, pressure tubes, and other components
for chemical facilities use stainless steel 304L and 316L [44]. In welding
joints, local heat treatment and micro-alloying can alter the mechanical
characteristics and microstructure of austenitic stainless steel [45]. Other
welding methods, such as explosive welding, can link austenitic stainless-
steel components to other materials, such as aluminum alloys [44]. Stainless
steel parts, however, may suffer from wear, fatigue, or damage after use.

Depending on the failure's complexity, size, and reason, the damaged components may need to be replaced or fixed. Various high-temperature material deposition techniques, like welding and laser metal deposition, are frequently used to repair steel components owing to their high interface, excellent component and deposited material quality, high operation reliability, and low cost [46, 47]. However, it could be challenging to implement these strategies in some circumstances due to the high heat impact. When complex geometric characteristics must be fixed, this impact might result in significant thermal stresses that distort the repaired portion. Recently, FSP has gained popularity due to its ability to make flawless high-quality welds and to combine incompatible materials without flaws. Perad et al. [44] tested the feasibility of cold spraying austenitic stainless steel on SS-304L using a tungsten carbide tool. The FSP treatment was carried out at a depth greater than the coating thickness using FSP parameters (traverse velocity 50 mm/min, rotational speed 300 rpm, axial force 20 kN, tilt angle 1.5°). Since it examines how FSP affects the structure and characteristics of cold spray coatings with high porosity and poor splat bonding, this study has been intriguing.

As-sprayed coatings were captured in cross-sectional micrographs at various magnifications (Figure 2.7(a–g)), no macroscale flaws or fractures were seen. However, the cross-sectional image at low magnification showed a sizable residual porosity (Figure 2.7(b)). The high porosity in the coating can be attributed to insufficient particle deformation during the cold spray procedure, which caused poor bonding between splats. In earlier research, Villa et al. [48], Sova et al. [49], and Adachi et al. [50] found that nitrogen-sprayed stainless steel cold spray deposits had significant porosity of up to 22%. The coating was composed of strongly bent stainless steel particles (splats) with unique surfaces, the typical structure for cold spray coatings. The FSP significantly changed the deposit system (Figure 2.7). Following FSP, a typical cross-sectional coating image revealed a change in the gray-scale that discriminated the substrate and coating material. Investigations conducted in cross-sections showed that the depth of the material affected by FSP was considerably greater than the coating thickness (Figure 2.7(e–g)). The FSP greatly decreased porosity, indicating that the method enhanced splat bonding. The considerable plasticization and deformation of the coating material during FSP, which developed a dense and fine-grained microstructure, resulted in better adhesion.

The microhardness map of as-sprayed coating showed non-uniform microhardness dispersion, with values ranging from less than 100 to 550 HV. The significant porosity in the coating was the cause of the hardness variations. The highest value of 598 HV, which shows particle work hardening after high-velocity impact and subsequent plastic deformation, was identified. Contrarily, the microhardness of the substrate was stable, measuring between 250 and 300 HV, comparable to the nanohardness of the substrate readings before the coating application. Following the FSP, the

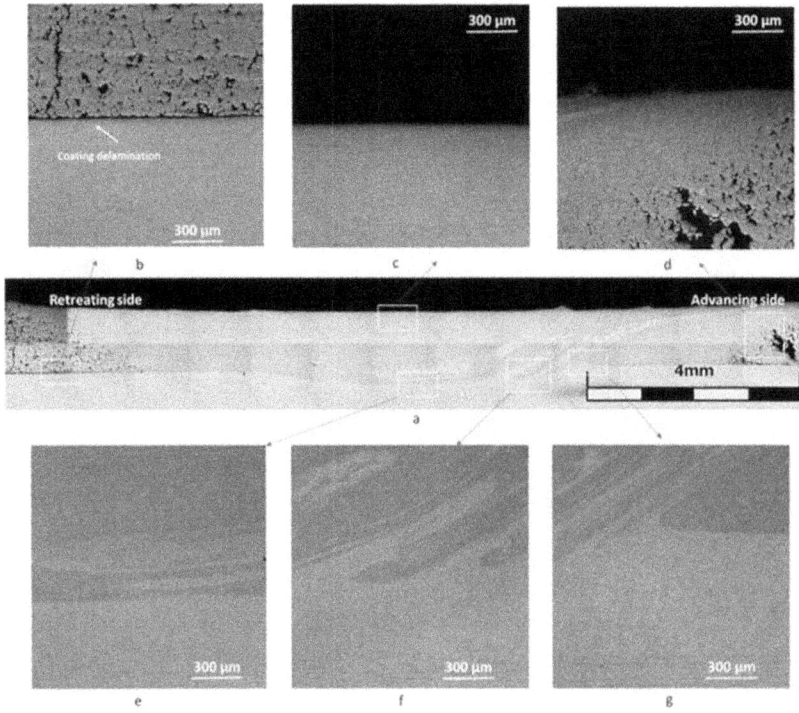

Figure 2.7 Sectional stereo micrographs of Diamalloy 1003 coatings on the 304L substrate after FSP as displayed: (a) board view; (b) failure of the adhesive at the coating's edge; (c) upper region of the coated surface; (d) cohesive failure at the coating perimeter; (e) surface–substrate interface at the stir zone; and (f, g) areas where substrate and coating materials mix [44].

deposit structure significantly changed, with microhardness homogenization in the affected zone, according to the microhardness map. After FSP, the microhardness map showed a decline in coating microhardness, with values averaging between 300 and 350 HV almost everywhere (Figure 2.8(b)). Peak values of 400–420 HV were observed, much less than the greatest hardness recorded in the as-sprayed deposit, where the substrate's substance changes to become the coating. Additionally, unaffected coated areas showed microhardness values that peaked at about 550 HV, which was higher.

The outcome showed that FSP could be a helpful technique for enhancing the characteristics of cold-sprayed coatings. The structure of the deposit was significantly affected by FSP, which caused microhardness uniformity in the damaged zone. The plastic deformation of the coated particles during FSP may be due to the homogeneity of the microhardness in the impacted zone. The reduced microhardness in the affected zone during FSP could be related to the eradication of poor bonding among the splats and lower porosity. The importance of cold-sprayed coating porosity control is also emphasized in

Microhardness HV$_{0.5}$

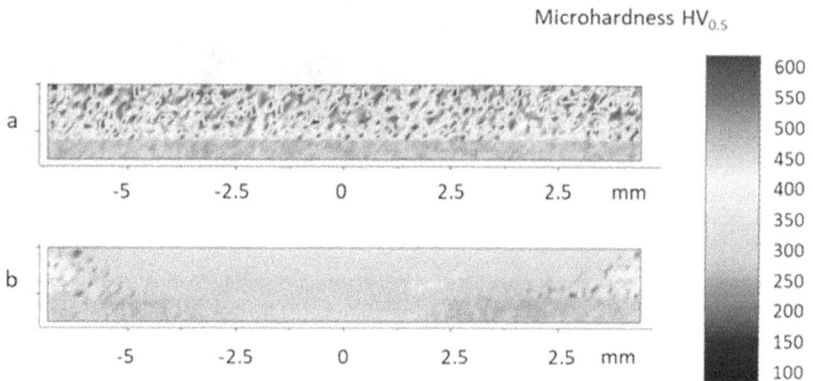

Figure 2.8 Microhardness of the 304L substrate's as-sprayed (a) FSP-treated (b) Diamalloy 1003 coatings [44].

the study. As can be noticed in the coating as it was sprayed, excessive porosity can result in a wide range of microhardness values. Better coating characteristics may arise from increasing the uniformity of the microhardness values by controlling the coating porosity during deposition. The outcome of this investigation indicates that FSP is a possible method for modifying the material characteristics and mechanical behavior of cold-sprayed coatings. This technique dramatically increased splat bonding while reducing porosity and residual stress. The FSP method also reduced the hardness of the coatings, bringing them closer to the substrates. Based on the study, where high-temperature deposition techniques like welding and laser fail, hybrid CS+FSP technology can be employed for repairing damaged stainless-steel components, especially those with intricate geometry.

2.4.3 Hybrid additive manufacturing of copper

Huang et al. [10] examined the mechanical behavior and morphological evaluation of cold-sprayed Cu60Zn40 alloy coatings on grit-blasted Cu plate substrates. The study concentrated on the CuZn alloy, a dual-phase transition system with non-equilibrium phases. The initial powder had equal-sized particles with a median size of 35.6 μm, and an irregular shape. The cold spray pistol with a MOC-style nozzle was operated by compressed high-pressure air at an operating temperature of 400 °C and with a pressure equivalent to 2.8 MPa. The coating was friction-stirred after being sprayed using an industry-standard FSW system with a traverse velocity of 100 mm/min and a tool rotation speed of 1500 rpm. The tool has a 2.5° (z-axis) slanted thread with a 3.4 mm diameter of the root and 1.5 mm long and a 10 mm concave shoulder. The study aimed to resolve several issues while enhancing the mechanical performance of the coating.

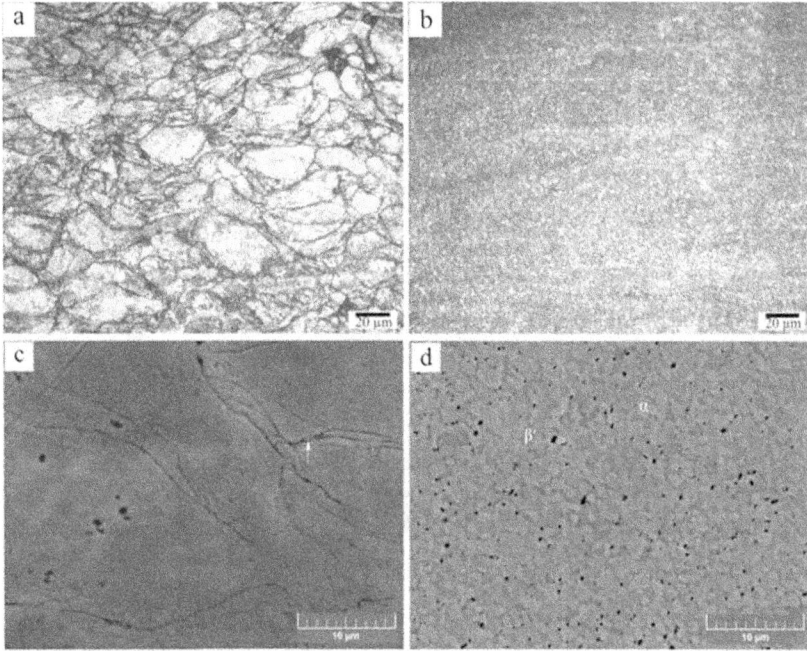

Figure 2.9 Sectional OM micrographs of (a) coatings that have been cold sprayed and (b) friction stirred, as well as higher magnification SEM micrographs of (c) cold sprayed and (d) friction stirred coatings [10].

The microstructure evolution after a cold-sprayed coating and a hybrid CS+FSP coating is shown in Figure 2.9(a–d). The planar structure of the cold-sprayed coating, with a median size of the grains being 1.9 μm and certain CuZn particles that have been stretched and twisted while leaving others undamaged, is depicted in Figure 2.9(a, c). Additionally, it is simple to see non-bonded particle-particle interfaces (white arrow), which indicates inhomogeneous deformation during deposition. On the other hand, the friction-stirred coating has an entirely recrystallized structure with 0.6 μm average grain size and ultrafine, equiaxed grains (Figure 2.9(b, d)). The bright gray "phase" and the dark gray "phase" are distributed within the grains. The thermo-mechanical coupled process produces complete recrystallization during FSP, comparable to severe plastic deformation (SPD) at high elevated temperatures [51]. Particularly for CuZn alloys with low stacking-fault energy (SFE), which obtain smaller grain sizes with SPD [52, 53], FSP is an excellent post-treatment for producing an ultrafine-grained coating for use in industry. The coating as it is sprayed may contain imperfections similar to large pores brought on by inadequate deformation, resulting in coatings with poor strength and excessive porosity. A thoroughly swirled coating with low porosity, excellent metallurgical bonding, and high strength is produced by FSP.

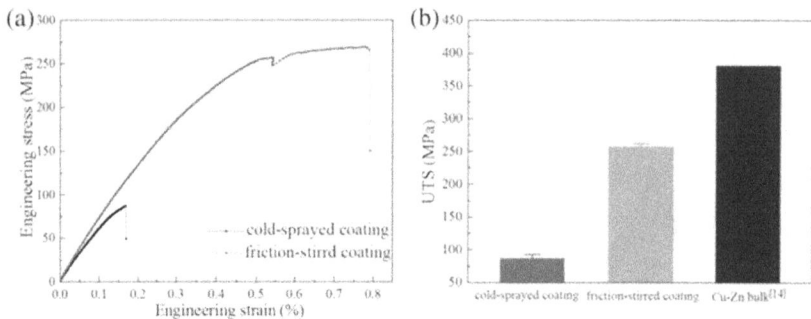

Figure 2.10 (a) Typical stress-strain plot and (b) ultimate tensile strength (UTS) of as-sprayed and friction-stirred coating [10].

The chemical and physical characteristics of the coatings as-sprayed and friction-stirred were analyzed, and it was found that they were made up of two primary phases, α and β'', before or after FSP processing. Further analysis of the coating following FSP indicates a larger crystallite size due to an increase in twins in the grains. With erroneous bars displaying a spectrum of test results for all three samples per situation, Figure 2.10 compares the tensile strength (UTS) of cold-sprayed and friction-stirred coatings. The cold-sprayed coating demonstrated brittleness with an average UTS of around 87.2 ± 5.9 MPa and less than 0.2% elongation to failure. Despite this, following FSP, the mean ultimate strength significantly increased, rising from the bulk-cast CuZn alloy's 381 MPa to 257.5 ± 4.8 MPa (Figure 2.10). The varied grain sizes, phase formulations, and fault dispersion may be responsible for the distinct fracture modes of the as-sprayed and friction-stirred coatings. Compared to the base metal, friction-stirred coatings exhibited a mean UTS of 257.5 ± 4.8 MPa and an elongation to failure of more than 0.8%. According to Zhang et al. [54], a drop in SFE and an increase in twinning capability may be advantageous to increase the ductility of friction-stirred coatings.

2.4.4 Composite material fabrication using hybrid additive technology

A metallic matrix that has been fortified by ceramic, metal, or organic particles makes up metal matrix composites (MMCs). MMCs have qualities not present in typical materials; these materials have properties that make them suitable for use in various aeronautical, automotive, and biomedical applications, including excellent specific strength, stiffness, and resistance to wear. Conventional MMC fabrication techniques like casting are not the best option for creating intricate shapes or specific properties. As a result, other production techniques, including hybrid CS+FSP technology, have been developed.

Complying with depositing both the metallic matrix and the reinforcing particles makes cold spray deposition a helpful technique for creating MMC coatings. In addition, it makes it possible to develop coatings that contain a lot of hard-reinforcing particles, which is difficult to do with conventional methods like casting. However, the reinforcing particle of MMC distribution may not be uniform. The particles may break up during the cold spray procedure, creating a non-homogeneous microstructure. FSP is preferred for making MMCs because it reduces porosity and redistributes the particles that reinforce the matrix for a more homogeneous microstructure.

Intending to construct an MMC in situ directly onto a plate's surface by creating a groove or slot on the plate containing reinforcement material, many studies have focused on producing metal matrix composites (MMCs) utilizing the FSP technique [55, 56]. However, particle ejection during FSP and non-uniform particle dispersion remain a problem. The cold spray approach has been proposed to place a layer of MMC on the surface of the plate, enhance material adherence to the surface, and provide a homogeneous coating microstructure in a single pass. Additionally, FSP can be employed after cold spraying to consolidate the material with a higher density. However, the influences on particle dispersion, morphology, and breakup during FSP in MMC coatings sprayed via cold spraying at reduced pressure conditions have not been extensively studied [57].

Additionally, there has never been a use of a simple cylindrical FSP tool. However, this might help to lower the costs associated with excessive tool wear when abrasion occurs during MMC material fabrication. Hodder et al. [9] investigated the morphology and behavior of the coating material and the effects of FSP and Al_2O_3 concentration on cold-sprayed Al-Al_2O_3 MMC coatings to address the general concerns. The study examined the feasibility of coating samples with an MMC coating comprising a significant amount (more than 40% by weight) of hard-reinforcing particles followed by FSP using inexpensive, simple cylindrical equipment. Before and following FSP, the deposition effectiveness, morphology, and hardness were investigated. Their findings demonstrate that the cold spray technique produces MMC coatings made of alumina and aluminum particles. Figure 2.11 schematically illustrates how FSP can result in a more equitable load distribution among the reinforcing particles during indentation.

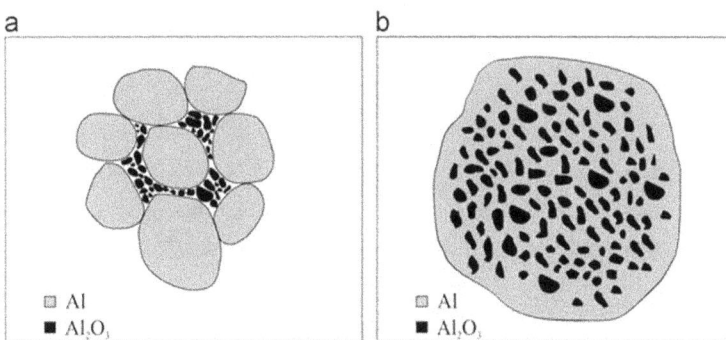

Figure 2.11 Diagram demonstrating the redistribution of reinforcing particles in MMC coatings after (a) CS and (b) FSP [9].

According to a study by Peat et al. [58], coatings such as aluminum oxide, tungsten carbide-cobalt chromium, and chromium carbide-nickel chromium successfully increased the hardness of AA5083 grade aluminum by about 540%. The coatings' hardness over the composite coating as it was deposited increased by 120%. It is significant to remember that even though FSP can enhance the microstructure and characteristics of MMC coatings, process parameter optimization is still required to obtain the desired outcomes. For instance, Yang et al. [59] looked into how the mechanical characteristics of cold-sprayed (CSed) AA2024/ Al_2O_3 (MMCs) were impacted by various rotating rpm during FSP. Because the refined Al_2O_3 particles have greater dispersion and stronger interparticle bonding, they found that raising the rotation speed to 1500 rpm allowed the UTS and elongation to increase by a maximum of 25.9% and 27.4%, respectively [60].

2.5 CORROSION OF HYBRID COMPOSITE MATERIAL

Yang et al. [11] investigated the potential for improving the ability to resist the corrosion of cold-sprayed AA2024/ Al_2O_3 MMCs by adopting FSP as an after-spray technique. In a solution of 3.5-weight percent NaCl, the electrochemical characteristics of the composites were examined using cyclic polarization. Figure 2.12 shows the common cyclic polarization curves of the CSed and FSPed AA2024/Al_2O_3 composites. Solid arrows indicate a possible scan direction. According to the forward scan, the CSed and FSPed stages created a protective layer in a passive area. Following the pitting potential

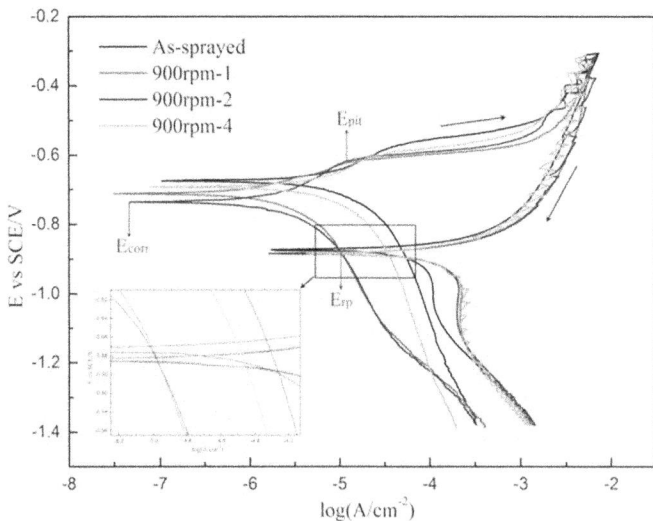

Figure 2.12 AA2024/Al2O3 coatings with CSed and FSPed cyclic polarization curves [11].

(Epit), the current density increased dramatically up to a point before steadily rising with potential. Recessivation with the repassivation potential (Erp) happened during the reverse scan. Since the significant conserved energy in the cold-sprayed and FSP MMCs, the negative hysteresis loops demonstrated limited capacity of the material to repassivate and unfavorable conditions for pitting formation. While Erp and Ecorr were related to the capacity for repassivation, The distinction among Epit and the corrosion potential (Ecorr) evaluated the propensity for pitting nucleation. Compared to the as-sprayed state, the (Epit - Ecorr) and (Erp - Ecorr) disparities after 1-pass and 2-pass FSP were more significant, indicating better-pitting resistance to corrosion and repassivity ability. The fact that Ecorr, Epit, and Erp's comparative current density values were lower under the 1- and 2-pass FSPed conditions than CSed gives credence to this idea. The corrosion achievement of the 2-pass FSP was superior to that of the 1-pass FSP, but it declined with the addition of a fourth FSP pass. The primary reason for boosting corrosion resistance is an improved surface condition, and, as a result, 2-pass FSP exhibits the best corrosion performance. The corrosion resistance, however, decreases with additional FSP passes (4 passes) resulting from the relationship between the better surface condition and degraded interior coating surfaces.

2.6 RESIDUAL STRESS IN HYBRID MANUFACTURING

This section aims to gain additional knowledge regarding the impact of the after-FSP on the residual stress distributions in the cold-sprayed coated layer. Major issues with the two-step surface combination of coating layers through the combined use of CS and FSP techniques and solid-state additive manufacturing have been discovered in earlier research [10, 61–64]. However, the intricate thermo-mechanical processing and mechanical restraint by workpiece gripping can result in thermal stresses that impact how much residual stress is left in the material after processing. Several papers have described the residual stress formation in various metals and alloys during FSP. On the other hand, numerous research has investigated the residual tension created on the subs-surface due to high-velocity particle collision during the CS deposition. Using the thermo-mechanical modeling techniques of Suhonen et al. [65] and Arabgol et al. [66], the generation and evolution of residual stress under cold spray coating of various metals were investigated.

An AA5083 alloy of the aluminum plate was chosen as the substrate in a study by Khodabakhshi et al. [61]. This research aims to determine how after-FSP changes impact residual stress distribution in cold-sprayed coating layers. This subject has never been the subject of investigation. A coating layer was produced using a commercially pure titanium powder on AA5083 aluminum alloy plate using a low-pressure condition cold spray approach. The titanium coating was deposited with CS and modified with FSP utilizing a Jafo milling machine using a tungsten carbide tool with no pins revolving clockwise. Utilizing a mixture of CS, FSP, and solid-state additive manufacturing, the study examined several systems and identified key issues related to the two-step surface integration of coating layers.

The residual stress pattern on the outer layer was measured utilizing (XRD) analysis both prior to and after post-CS deposition and after FSP alteration. Using Bruker, Inc D8-Advance X-ray diffractometer equipped with a VNTEC-500 area detection (radius of 135 mm), the experiments were carried out entirely on the surface, with an interaction volume depth of approximately 5 μm. For the titanium coating and the aluminum alloy of 5083 series as a substrate, respectively, the two angles among the incident beam and the detector, with a collimator size of around 1.0 mm, were consistent throughout each experiment at approximately 99.22° and 109.01°.

The instrument was mounted on a stage with motors and fluctuated with amplitudes of 1.5 and 2.5 mm between the x- and y-axis and speeds of 3.5 and 5.5 mm/s to cover the Ψ-scan in a tilting between 0° and 50° with an angle step size of 25°. The range of tilting angles for the Φ-scan, which was used to specify sample rotation, was increased from 0° to 360° with a 45° angle step. The plane to be used to test residual stress for titanium coating was given the designation (211). The area detector recorded the Debye-Scherrer diffraction bands and produced a two-dimensional image. The signa x and y represent the measurements along longitudinal and transverse directions.

Figure 2.13(a) at three separate points, A, B, and C, displays the divergent residual stress values across the rolling direction of the original AA5083 alloy substrate for both longitudinal (X) and transverse (Y) directions. These observations show that the surface of the initial sheet underwent thermomechanical processing, resulting in a generally compressive residual stress between 20 and 45 MPa. Figure 2.13(b) depicts longitudinal and transversal residual stress variations on the surface of titanium-coated layer after CS deposition. The measurements demonstrate that the residual stress is still compressive following the high-speed impact of the titanium particles, with the greatest value produced by particle impingement being roughly 50 MPa in the longitudinal direction. Additionally, given the one-by-one particle deposition method used during CS, the data shows a sizable divergence in the trend of residual stress across the x and y planes. Still, directionality of this trend is not primarily due to CS deposition but rather to discrete thermal contractions of the substrate along the x- and y-axes.

The distribution of the residual stress of titanium coatings after FSP varied with high and low plunge depths is shown in Figure 2.14(a–d) in both longitudinal and transverse directions. Relative stress patterns for samples with high and low plunge depths varied significantly. The asymmetric change of residual stress distribution inside the stir zone can be linked to the asymmetric character of FSP [1, 67–72]. The residual stress distribution becomes more compressive when the plunge depth is increased because of the higher volumetric heat input, which improves interaction with the tool shoulder and sheet surface and increases material density and flow on the outer layer (Figure 2.14(a, b)) [73]. The transverse residual stress distribution is less negative (Figure 2.14(c, d)). It gradually changes to more compressive values

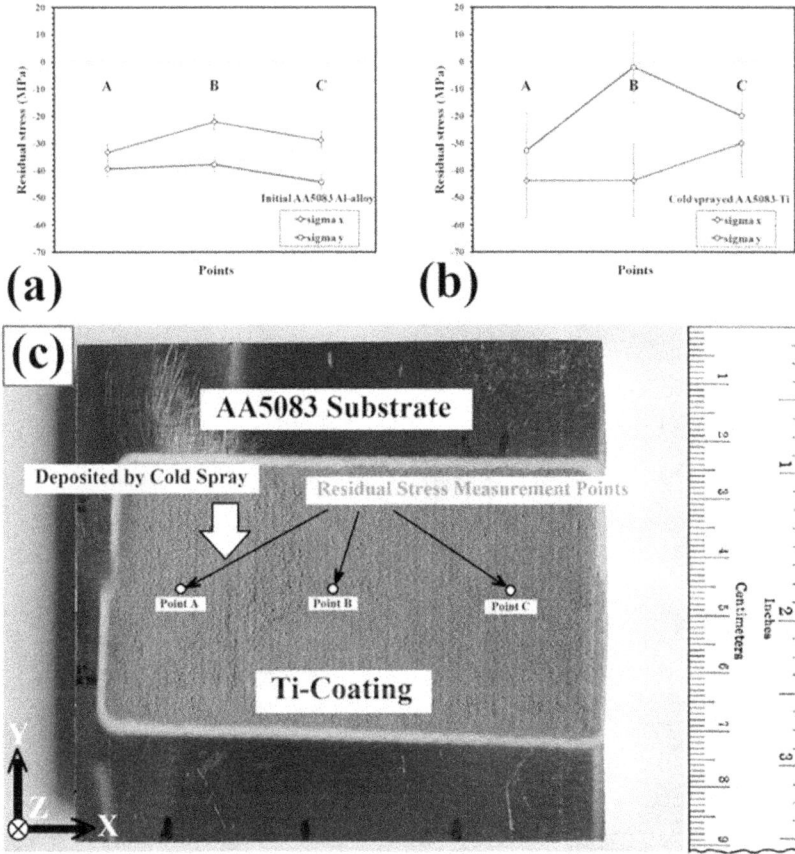

Figure 2.13 Residual distributions of stress in the x- and y-axis on the surfaces of (a) an aluminum substrate, (b) cold gas-sprayed titanium covering at three distinct positions on the sheet-rolling direction designated by the letters A, B, and C, and (c) a view of the titanium coating layer at a macro scale [61].

in the direction of the moving and retreating sides in samples with a large plunge distance (Figure 2.14(c)). Low plunge depth samples exhibit tensile character in a few areas (Figure 2.14(b, d)).

A deeper plunge depth substantially enhances the residual stress profiles, which also show a general compressive trend in relation to the FSP plunge depth and transverse location. The shoulder friction-stirring effect may be to blame for the advancing side achieving the highest compressive residual stress value of about 400 MPa. It's also crucial to keep in mind that using a larger plunge depth increases the consistency of the stress residue distribution throughout the length of the modified sample as opposed to using a shallower plunge depth, which results in a modified sample with less uniformity/homogeneity.

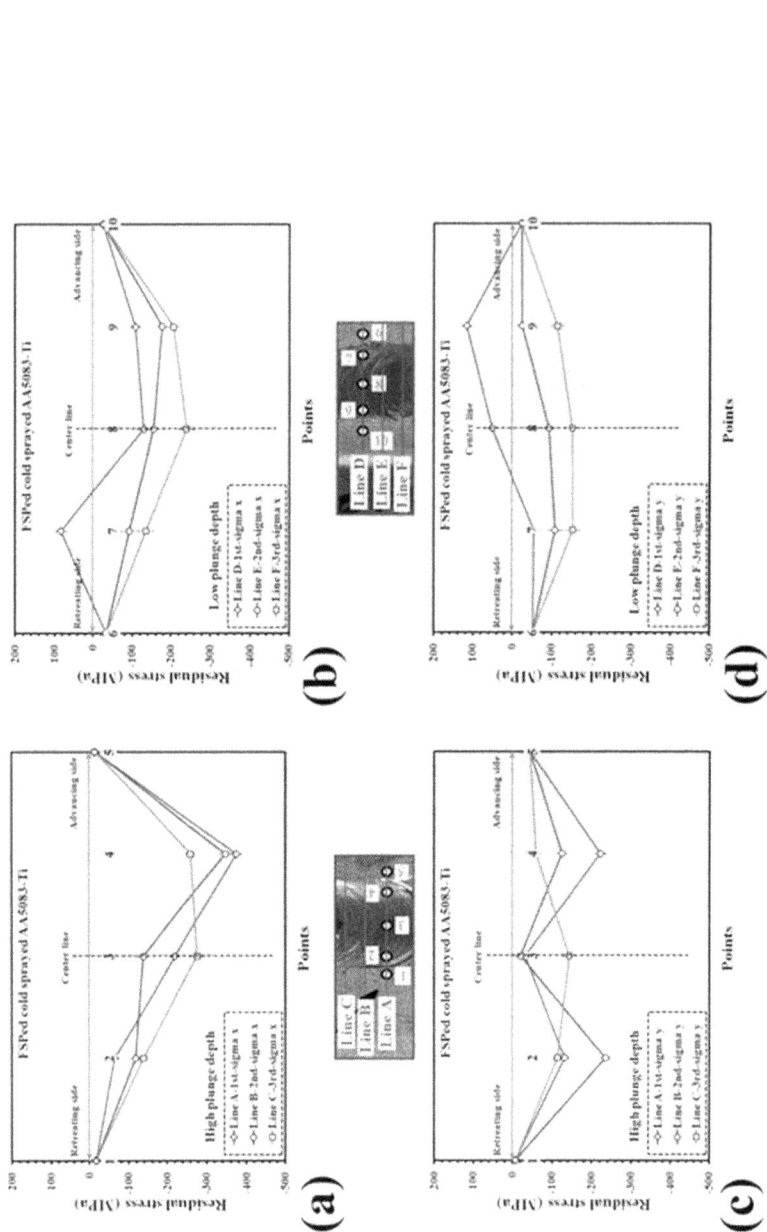

Figure 2.14 Surface residual stress profiles of a friction-stir modified titanium coating in the (a, b) x- and (c, d) y-directions at (a, c) high and (b, d) low plunge depths for three distinct lines across the transverse section are shown in Figure 2.1b [61].

2.7 APPLICATIONS, CHALLENGES, AND FUTURE DIRECTIONS

Application possibilities for the CS and FSP combination in hybrid additive manufacturing include the aerospace, automotive, and medical device sectors. Production of graded materials, refinement in microstructural for end-product customization, composite production, and the creation of stiffened structures are just a few examples of potential applications [1, 29, 74, 75]. Before commercialization, however, a number of issues related to hybrid CS+FSP must be resolved. For instance, manufacturing elaborate structures necessitates specialized fittings, in contrast to fusion-based AM methods, which allow for the creation of complicated net-shaped structures. The small feature size that can be printed with hybrid CS+FSP, which is currently 10 mm [76], can also be increased by lowering the size of the feed and tooling materials. Another issue with the components is their mechanical stability since, when made using hybrid CS+FSP, high aspect ratio components might occasionally buckle from two-step heat input. To account for a gradual shift at the edge of each layer, the post-processing machining process of the margins is also necessary.

Furthermore, the capital expense is problematic because complex forms must be fabricated using specialized clamps and specially designed fixtures, raising the price of components. The hybrid CS+FSP additive manufacturing technique depends on the ability of the tool to stir the material, and tool health and lifetime are significant determinants of performance. High-strength, wear- and fatigue-resistant tool materials are required for working with high-strength materials. The tool currently prevents the application of hybrid CS+FSP to high-temperature materials. Although tools made of PCBN and W-Re work well, their higher cost and proper process parameter management are required for component sustainability. In soft materials, cold spray can efficiently disperse ceramic particles, reducing tool wear and failure brought on by uniform material flow, shear force, and torque. But tool wear is inevitable during the FSP of ceramic-based composites and elevated temperatures materials.

The absence of quantifiable correlations between process controls and the final microstructure and characteristics poses a significant obstacle to developing hybrid CS+FSP additive manufacturing techniques. This is done so that the forces and temperatures generated during FSP cannot be monitored by tracking with control via feedback. A fully instrumented machine can obtain real-time feedback for process control and tracking, machine and tool health monitoring using in situ sensors like thermal imaging and force and torque sensors. Researchers must tackle these technical issues since hybrid CS+FSP additive manufacturing is currently in its early phases if we are to comprehend these revolutionary techniques better. Developing friction-stir-based additive manufacturing processes and innovative solutions to the aforementioned constraints will significantly broaden research from conception through component-level design and manufacture.

REFERENCES

[1] S. Kumar, A. Kar, A Review of Solid-State Additive Manufacturing Processes, *Transactions of the Indian National Academy of Engineering*, 6 (2021) 955–973.

[2] N. Ur Rahman, L. Capuano, S. Cabeza, M. Feinaeugle, A. Garcia-Junceda, M.B. de Rooij, D.T.A. Matthews, G. Walmag, I. Gibson, G.R.B.E. Römer, Directed energy deposition and characterization of high-carbon high speed steels, *Additive Manufacturing*, 30 (2019) 100838.

[3] A. Papyrin, V. Kosarev, S. Klinkov, A. Alkimov, V. Fomin, Chapter 1 – Discovery of the cold spray phenomenon and its basic features, in: A. Papyrin, V. Kosarev, S. Klinkov, A. Alkimov, V. Fomin (Eds.) *Cold Spray Technology*, Elsevier, Oxford, 2007, pp. 1–32.

[4] A. Leon, G. Katarivas Levy, T. Ron, A. Shirizly, E. Aghion, The effect of strain rate on stress corrosion performance of Ti6Al4V alloy produced by additive manufacturing process, *Journal of Materials Research and Technology*, 9 (2020) 4097–4105.

[5] W.-Y. Li, H. Liao, C.-J. Li, G. Li, C. Coddet, X. Wang, On high velocity impact of micro-sized metallic particles in cold spraying, *Applied Surface Science*, 253 (2006) 2852–2862.

[6] P. Cavaliere, A. Silvello, Fatigue behaviour of cold sprayed metals and alloys: critical review, *Surface Engineering*, 32 (2016) 631–640.

[7] N. Deng, J. Tang, T. Xiong, J. Li, Z. Zhou, Fabrication and characterization of WCu composite coatings with different W contents by cold spraying, *Surface and Coatings Technology*, 368 (2019) 8–14.

[8] M.R. Rokni, C.A. Widener, V.K. Champagne, G.A. Crawford, S.R. Nutt, The effects of heat treatment on 7075 Al cold spray deposits, *Surface and Coatings Technology*, 310 (2017) 278–285.

[9] K.J. Hodder, H. Izadi, A.G. McDonald, A.P. Gerlich, Fabrication of aluminum–alumina metal matrix composites via cold gas dynamic spraying at low pressure followed by friction stir processing, *Materials Science and Engineering: A*, 556 (2012) 114–121.

[10] C. Huang, W. Li, Y. Feng, Y. Xie, M.-P. Planche, H. Liao, G. Montavon, Microstructural evolution and mechanical properties enhancement of a cold-sprayed CuZn alloy coating with friction stir processing, *Materials Characterization*, 125 (2017) 76–82.

[11] K. Yang, W. Li, Y. Xu, X. Yang, Using friction stir processing to augment corrosion resistance of cold sprayed AA2024/Al2O3 composite coatings, *Journal of Alloys and Compounds*, 774 (2019) 1223–1232.

[12] W. Li, C. Cao, G. Wang, F. Wang, Y. Xu, X. Yang, 'Cold spray +' as a new hybrid additive manufacturing technology: a literature review, *Science and Technology of Welding and Joining*, 24 (2019) 420–445.

[13] C. Chen, Y. Xie, S. Yin, W. Li, X. Luo, X. Xie, R. Zhao, C. Deng, J. Wang, H. Liao, M. Liu, Z. Ren, Ductile and high strength Cu fabricated by solid-state cold spray additive manufacturing, *Journal of Materials Science & Technology*, 134 (2023) 234–243.

[14] M. Oksa, E. Turunen, T. Suhonen, T. Varis, S.-P. Hannula, Optimization and characterization of high velocity oxy-fuel sprayed coatings: techniques, materials, and applications, *Coatings*, 1 (2011) 17–52.

[15] P. Fauchais, A. Vardelle, M. Vardelle, M. Fukumoto, Knowledge concerning splat formation: an invited review, *Journal of Thermal Spray Technology*, 13 (2004) 337–360.

[16] A. Alkhimov, V. Kosarev, A. Papyrin, A method of cold gas dynamic deposition, Dokl. Akad. Nauk. SSSR 315 (5), (1990).

[17] B.L. James, V.S. Bhattiprolu, G.A. Crawford, B.K. Jasthi, Effect of zirconia secondary peening on the microstructure and mechanical behavior of Al6061 cold spray coatings, *Surface and Coatings Technology*, 436 (2022) 128–269.

[18] A. Papyrin, Cold spray technology, *Advanced Materials & Processes*, 159 (2001) 49–49.

[19] S. Yin, P. Cavaliere, B. Aldwell, R. Jenkins, H. Liao, W. Li, R. Lupoi, Cold spray additive manufacturing and repair: Fundamentals and applications, *Additive Manufacturing*, 21 (2018) 628–650.

[20] H. Assadi, H. Kreye, F. Gärtner, T. Klassen, Cold spraying – A materials perspective, *Acta Materialia*, 116 (2016) 382–407.

[21] V.S. Bhattiprolu, K.W. Johnson, O.C. Ozdemir, G.A. Crawford, Influence of feedstock powder and cold spray processing parameters on microstructure and mechanical properties of Ti-6Al-4V cold spray depositions, *Surface and Coatings Technology*, 335 (2018) 1–12.

[22] T. Hussain, D.G. McCartney, P.H. Shipway, D. Zhang, Bonding mechanisms in cold spraying: The contributions of metallurgical and mechanical components, *Journal of Thermal Spray Technology*, 18 (2009) 364–379.

[23] S. Nemati, L.G. Butler, K. Ham, G.L. Knapp, C. Zeng, S. Emanet, H. Ghadimi, S. Guo, Y. Zhang, H. Bilheux, Neutron imaging of Al6061 prepared by solid-state friction stir additive manufacturing, *Metals*, 13 (2023) 188.

[24] D. Shah, V.J. Badheka, Friction stir additive manufacturing—A review, in: A.K. Parwani, P.L. Ramkumar, K. Abhishek, S.K. Yadav (Eds.) *Recent Advances in Mechanical Infrastructure*, Springer Singapore, Singapore, 2021, pp. 13–36.

[25] A. Vardelle, C. Moreau, J. Akedo, H. Ashrafizadeh, C.C. Berndt, J.O. Berghaus, M. Boulos, J. Brogan, A.C. Bourtsalas, A. Dolatabadi, The 2016 thermal spray roadmap, *Journal of Thermal Spray Technology*, 25 (2016) 1376–1440.

[26] S. Pathak, G.C. Saha, Development of sustainable cold spray coatings and 3D additive manufacturing components for repair/manufacturing applications: A critical review, *Coatings*, 7 (2017) 122.

[27] C. Li, W. Li, Y. Wang, H. Fukanuma, Effect of spray angle on deposition characteristics in cold spraying, *Thermal Spray*, (2003) 91–96.

[28] F. Gärtner, T. Stoltenhoff, T. Schmidt, H. Kreye, The cold spray process and its potential for industrial applications, *Journal of Thermal Spray Technology*, 15 (2006) 223–232.

[29] R.S. Mishra, R.S. Haridas, P. Agrawal, Friction stir-based additive manufacturing, *Science and Technology of Welding and Joining*, 27 (2022) 141–165.

[30] R. Kumar, S. Chattopadhyaya, A.R. Dixit, B. Bora, M. Zelenak, J. Foldyna, S. Hloch, P. Hlavacek, J. Scucka, J. Klich, L. Sitek, P. Vilaca, Surface integrity analysis of abrasive water jet-cut surfaces of friction stir welded joints, *The International Journal of Advanced Manufacturing Technology*, 88 (2017) 1687–1701.

[31] A. Kumar Srivastava, N. Kumar, A. Rai Dixit, Friction stir additive manufacturing – An innovative tool to enhance mechanical and microstructural properties, *Materials Science and Engineering: B*, 263 (2021) 114832.

[32] W. Wang, P. Han, Y. Wang, T. Zhang, P. Peng, K. Qiao, Z. Wang, Z. Liu, K. Wang, High-performance bulk pure Al prepared through cold spray-friction stir processing composite additive manufacturing, *Journal of Materials Research and Technology*, 9 (2020) 9073–9079.

[33] J. Pattison, S. Celotto, R. Morgan, M. Bray, W. O'Neill, Cold gas dynamic manufacturing: A non-thermal approach to freeform fabrication, *International Journal of Machine Tools and Manufacture*, 47 (2007) 627–634.

[34] J.M. Flynn, A. Shokrani, S.T. Newman, V. Dhokia, Hybrid additive and subtractive machine tools – Research and industrial developments, *International Journal of Machine Tools and Manufacture*, 101 (2016) 79–101.

[35] S. Cadney, M. Brochu, P. Richer, B. Jodoin, Cold gas dynamic spraying as a method for freeforming and joining materials, *Surface and Coatings Technology*, 202 (2008) 2801–2806.

[36] Y.-H. Ho, S.S. Joshi, T.-C. Wu, C.-M. Hung, N.-J. Ho, N.B. Dahotre, In-vitro bio-corrosion behavior of friction stir additively manufactured AZ31B magnesium alloy-hydroxyapatite composites, *Materials Science and Engineering: C*, 109 (2020) 110632.

[37] J.J.S. Dilip, G.D. Janaki Ram, B.E. Stucker, Additive manufacturing with friction welding and friction deposition processes, *International Journal of Rapid Manufacturing*, 3 (2012) 56–69.

[38] Q. Han, R. Setchi, F. Lacan, D. Gu, S.L. Evans, Selective laser melting of advanced Al-Al2O3 nanocomposites: Simulation, microstructure and mechanical properties, *Materials Science and Engineering: A*, 698 (2017) 162–173.

[39] M.P. Reddy, V. Manakari, G. Parande, F. Ubaid, R. Shakoor, A. Mohamed, M. Gupta, Enhancing compressive, tensile, thermal and damping response of pure Al using BN nanoparticles, *Journal of Alloys and Compounds*, 762 (2018) 398–408.

[40] M.I. Abd El Aal, M. Sadawy, Influence of ECAP as grain refinement technique on microstructure evolution, mechanical properties and corrosion behavior of pure aluminum, *Transactions of Nonferrous Metals Society of China*, 25 (2015) 3865–3876.

[41] N. Read, W. Wang, K. Essa, M.M. Attallah, Selective laser melting of AlSi10Mg alloy: Process optimisation and mechanical properties development, *Materials & Design* (1980-2015), 65 (2015) 417–424.

[42] F. Khodabakhshi, B. Marzbanrad, L.H. Shah, H. Jahed, A.P. Gerlich, Friction-stir processing of a cold sprayed AA7075 coating layer on the AZ31B substrate: Structural homogeneity, microstructures and hardness, *Surface and Coatings Technology*, 331 (2017) 116–128.

[43] G. Ji, H. Liu, G.-J. Yang, C.-X. Li, X.-T. Luo, G.-Y. He, L. Zhou, Effect of friction stir spot processing on microstructure and mechanical properties of cold-sprayed Al coating on Ti substrate, *Surface and Coatings Technology*, 421 (2021) 127352.

[44] T. Perard, A. Sova, H. Robe, V. Robin, Y. Zedan, P. Bocher, E. Feulvarch, Friction stir processing of austenitic stainless steel cold spray coating deposited on 304L stainless steel substrate: feasibility study, *The International Journal of Advanced Manufacturing Technology*, 115 (2021) 2379–2393.

[45] G. Khalaj, H. Pouraliakbar, M.R. Jandaghi, A. Gholami, Microalloyed steel welds by HF-ERW technique: Novel PWHT cycles, microstructure evolution and mechanical properties enhancement, *International Journal of Pressure Vessels and Piping*, 152 (2017) 15–26.

[46] C. Gunter, M.P. Miles, F.C. Liu, T.W. Nelson, Solid state crack repair by friction stir processing in 304L stainless steel, *Journal of Materials Science & Technology*, 34 (2018) 140–147.

[47] X. Lin, Y. Cao, X. Wu, H. Yang, J. Chen, W. Huang, Microstructure and mechanical properties of laser forming repaired 17-4PH stainless steel, *Materials Science and Engineering: A*, 553 (2012) 80–88.

[48] M. Villa, S. Dosta, J.M. Guilemany, Optimization of 316L stainless steel coatings on light alloys using Cold Gas Spray, *Surface and Coatings Technology*, 235 (2013) 220–225.

[49] A. Sova, S. Grigoriev, A. Okunkova, I. Smurov, Cold spray deposition of 316L stainless steel coatings on aluminium surface with following laser post-treatment, *Surface and Coatings Technology*, 235 (2013) 283–289.

[50] S. Adachi, N. Ueda, Effect of cold-spray conditions using a nitrogen propellant gas on AISI 316L stainless steel-coating microstructures, *Coatings*, 7 (2017) 87.

[51] Z.H. Zhang, W.Y. Li, Y. Feng, J.L. Li, Y.J. Chao, Global anisotropic response of friction stir welded 2024 aluminum sheets, *Acta Materialia*, 92 (2015) 117–125.

[52] X.L. Ma, H. Zhou, J. Narayan, Y.T. Zhu, Stacking-fault energy effect on zero-strain deformation twinning in nanocrystalline Cu–Zn alloys, *Scripta Materialia*, 109 (2015) 89–93.

[53] Y.H. Zhao, Z. Horita, T.G. Langdon, Y.T. Zhu, Evolution of defect structures during cold rolling of ultrafine-grained Cu and Cu–Zn alloys: Influence of stacking fault energy, *Materials Science and Engineering: A*, 474 (2008) 342–347.

[54] P. Zhang, X.H. An, Z.J. Zhang, S.D. Wu, S.X. Li, Z.F. Zhang, R.B. Figueiredo, N. Gao, T.G. Langdon, Optimizing strength and ductility of Cu–Zn alloys through severe plastic deformation, *Scripta Materialia*, 67 (2012) 871–874.

[55] Z. Ma, Friction stir processing technology: a review, *Metallurgical and materials Transactions A*, 39 (2008) 642–658.

[56] H.C. Madhu, V. Edachery, K.P. Lijesh, C.S. Perugu, S.V. Kailas, Fabrication of wear-resistant Ti3AlC2/Al3Ti hybrid aluminum composites by friction stir processing, *Metallurgical and Materials Transactions A*, 51 (2020) 4086–4099.

[57] A. Pirondi, L. Collini, Analysis of crack propagation resistance of Al–Al2O3 particulate-reinforced composite friction stir welded butt joints, *International Journal of Fatigue*, 31 (2009) 111–121.

[58] T. Peat, A. Galloway, A. Toumpis, P. McNutt, N. Iqbal, The erosion performance of cold spray deposited metal matrix composite coatings with subsequent friction stir processing, *Applied Surface Science*, 396 (2017) 1635–1648.

[59] K. Yang, W. Li, C. Huang, X. Yang, Y. Xu, Optimization of cold-sprayed AA2024/Al2O3 metal matrix composites via friction stir processing: Effect of rotation speeds, *Journal of Materials Science & Technology*, 34 (2018) 2167–2177.

[60] K. Yang, W. Li, P. Niu, X. Yang, Y. Xu, Cold sprayed AA2024/Al2O3 metal matrix composites improved by friction stir processing: Microstructure characterization, mechanical performance and strengthening mechanisms, *Journal of Alloys and Compounds*, 736 (2018) 115–123.

[61] F. Khodabakhshi, B. Marzbanrad, A. Yazdanmehr, H. Jahed, A.P. Gerlich, Tailoring the residual stress during two-step cold gas spraying and friction-stir surface integration of titanium coating, *Surface and Coatings Technology*, 380 (2019) 125008.

[62] F. Khodabakhshi, B. Marzbanrad, H. Jahed, A.P. Gerlich, Interfacial bonding mechanisms between aluminum and titanium during cold gas spraying followed by friction-stir modification, *Applied Surface Science*, 462 (2018) 739–752.

[63] F. Khodabakhshi, A.P. Gerlich, Potentials and strategies of solid-state additive friction-stir manufacturing technology: A critical review, *Journal of Manufacturing Processes*, 36 (2018) 77–92.

[64] T. Peat, A. Galloway, A. Toumpis, R. Steel, W. Zhu, N. Iqbal, Enhanced erosion performance of cold spray co-deposited AISI316 MMCs modified by friction stir processing, *Materials & Design*, 120 (2017) 22–35.

[65] T. Suhonen, T. Varis, S. Dosta, M. Torrell, J.M. Guilemany, Residual stress development in cold sprayed Al, Cu and Ti coatings, *Acta Materialia*, 61 (2013) 6329–6337.

[66] Z. Arabgol, H. Assadi, T. Schmidt, F. Gärtner, T. Klassen, Analysis of thermal history and residual stress in Cold-Sprayed Coatings, *Journal of Thermal Spray Technology*, 23 (2014) 84–90.

[67] A. Kar, D. Yadav, S. Suwas, S.V. Kailas, Role of plastic deformation mechanisms during the microstructural evolution and intermetallics formation in dissimilar friction stir weld, *Materials Characterization*, 164 (2020) 110371.

[68] A. Kar, S.K. Choudhury, S. Suwas, S.V. Kailas, Effect of niobium interlayer in dissimilar friction stir welding of aluminum to titanium, *Materials Characterization*, 145 (2018) 402–412.

[69] A. Kar, T. Curtis, B.K. Jasthi, W. Lein, Z. McClelland, G. Crawford, Mechanism of joint formation in dissimilar friction stir welding of Aluminum to Steel, in: Y. Hovanski, Y. Sato, P. Upadhyay, A.A. Naumov, N. Kumar (Eds.) *Friction Stir Welding and Processing XII*, Springer Nature Switzerland, Cham, 2023, pp. 237–245.

[70] A. Kar, S. Suwas, S.V. Kailas, Multi-Length Scale Characterization of Microstructure Evolution and Its Consequence on Mechanical Properties in Dissimilar Friction Stir Welding of Titanium to Aluminum, *Metallurgical and Materials Transactions A*, 50 (2019) 5153–5173.

[71] A. Kar, B. Vicharapu, Y. Morisada, H. Fujii, Elucidation of interfacial microstructure and properties in friction stir lap welding of aluminium alloy and mild steel, *Materials Characterization*, 168 (2020) 110572.

[72] H. Jamshidi Aval, Microstructure and residual stress distributions in friction stir welding of dissimilar aluminium alloys, *Materials & Design*, 87 (2015) 405–413.

[73] A. Shamsipur, S.F. Kashani-Bozorg, A. Zarei-Hanzaki, Production of in-situ hard Ti/TiN composite surface layers on CP-Ti using reactive friction stir processing under nitrogen environment, *Surface and Coatings Technology*, 218 (2013) 62–70.

[74] M. Yuqing, K. Liming, H. Chunping, L. Fencheng, L. Qiang, Formation characteristic, microstructure, and mechanical performances of aluminum-based components by friction stir additive manufacturing, *The International Journal of Advanced Manufacturing Technology*, 83 (2016) 1637–1647.

[75] S. Palanivel, H. Sidhar, R.S. Mishra, Friction stir additive manufacturing: Route to high structural performance, *JOM*, 67 (2015) 616–621.

[76] H.Z. Yu, M.E. Jones, G.W. Brady, R.J. Griffiths, D. Garcia, H.A. Rauch, C.D. Cox, N. Hardwick, Non-beam-based metal additive manufacturing enabled by additive friction stir deposition, *Scripta Materialia*, 153 (2018) 122–130.

Chapter 3

Additive friction stir deposition

Vipin Gopan
St. Thomas College of Engineering and Technology, Alappuzha, India

Valliyappan David Natarajan
Universiti Teknologi MARA, Shah Alam, Malaysia

S. Sarath and Roshin Thomas Varughese
St. Thomas College of Engineering and Technology, Alappuzha, India

Praveen Kumar
Mahaguru Institute of Technology, Alappuzha, India

3.1 INTRODUCTION

The fourth industrial revolution, comprising advanced digital technologies including the Internet of Things (IoT), artificial intelligence (AI), Big Data analytics, augmented reality, etc., has witnessed a paradigm shift in the functioning of contemporary manufacturing industries [1–3]. The role of additive manufacturing, a key element of the fourth industrial revolution, is exponentially increasing in modern manufacturing and market scenarios where customized parts/products and streamlined supply chains are of considerable interest [4, 5]. According to the American Society for Testing and Materials (ASTM), additive manufacturing can be categorized into seven distinct techniques [6]. However, in the case of metal additive manufacturing, fusion-based or beam-based technologies, including powder bed fusion (PBF) and directed energy deposition (DED), are the predominant techniques [7–9]. These techniques use lasers or electron beams for the melting and consolidation of metal powders. Though fusion-based additive manufacturing technologies offer a multitude of advantages, including the availability of a wide range of materials, extreme geometric complexities, the ability to print FGM, metal powder recycling, etc., there are certain inherent constraints that restrict its widespread application [10, 11]. Inability to control the porosity (keyhole and lack of fusion porosity) and anisotropic residual stress, columnar grain structure restricting the microstructure control, expensive and time-consuming powder preparation, spatter ejection, large thermal gradients, large energy requirements, need for post-processing, etc., still remain the fundamental challenges to the process despite the extensive research and advancements made in recent years [12–14].

DOI: 10.1201/9781032616025-3

To troubleshoot these fundamental issues, in recent years, much research has been directed towards developing solid-state additive manufacturing techniques that utilize plastic deformation, rather than melting and solidification, for metal additive manufacturing that can altogether open new avenues. These include cold spray additive manufacturing, friction deposition, ultrasonic additive manufacturing, friction surfacing, friction stir additive manufacturing, additive friction stir deposition, etc. [12, 15–17].

Additive friction stir deposition, deriving from the fundamental principles of friction stir welding/processing (FSW/P), is the only solid-state additive manufacturing technology capable of producing fully dense, equiaxed, wrought-like microstructures and mechanical properties in the as-printed state. AFSD has immense potential in terms of defense, aerospace, and automotive applications as it is the forging counterpart of laser-based AM processes. Though recent years witnessed several research activities in the AFSD domain, the underlying fundamental principles are yet to be fully understood. This chapter summarizes the basic principles, features, and capabilities of AFSD, the critical findings in this domain with particular emphasis on the investigation of microstructural and mechanical properties, its present limitations, and then examines the future scope of development in this domain.

3.2 WORKING PRINCIPLE

Although several other friction-based additive manufacturing techniques are available, AFSD is simply a layer-by-layer building additive process that results in the production of near net-shaped components. In the AFSD process, the material to be deposited, in the form of either rod or powder, is fed through a rotating hollow shoulder which is the AFSD tool. As the material is fed through the AFSD tool, high frictional heat is generated at the contact surfaces, i.e., between the material and tool and between the material and substrate. Owing to the high rise in temperature of the feed material, it becomes plasticized and softened, and extrudes to fill the gap between the rotating tool and the substrate [15]. The rotating bottom of the tool serves two purposes: (i) controlling the thickness of the deposited layer and (ii) providing shearing to the deposited material, which ensures good deposition with minimum porosity. After depositing the first layer, the tool moves upwards by the thickness of one layer and the subsequent layers are printed on top of the initial one. At first glance, the AFSD principle looks similar to fused deposition modeling, which simply applies heating and in which the in-plane motion of the tool relative to the substrate helps in achieving the layer-by-layer printing of 3D parts. However, AFSD relies heavily on the friction stir principle, where the plastic deformation of the feed material results in good interface bonding between the substrate and the deposited material [18–20]. This technology was first commercialized and patented by Aeroprobe Corporation [21]. The schematic diagram of the AFSD process is illustrated in Figure 3.1.

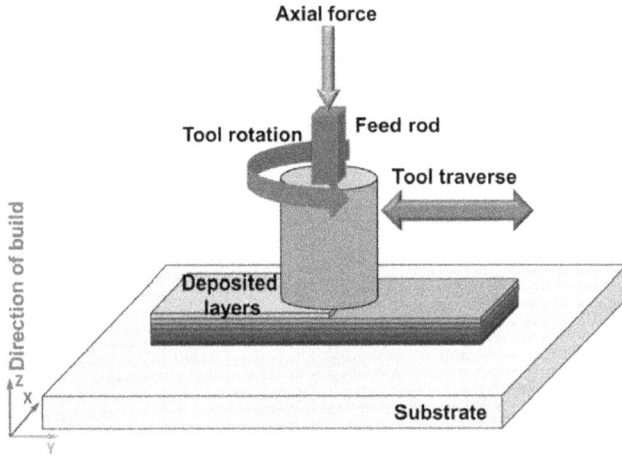

Figure 3.1 Additive friction stir deposition process [15].

3.3 SALIENT FEATURES OF AFSD

Being a thermomechanical process and the material addition occurring as a result of the severe plastic deformation at high temperatures, AFSD offers inherent advantages over fusion-based additive manufacturing techniques.

3.3.1 Porosity and void formation

One of the inherent limitations of the fusion-based additive manufacturing techniques is the porosity, which can be categorized into two distinct types: keyhole porosity and lack of fusion porosity. The incomplete melting of powders contributes mainly to the lack of fusion porosity, whereas high energy input causing the vaporization of metals results in the entrapment of gas bubbles beneath the surface resulting in keyhole porosity [22–24]. Another type of porosity which occurs in the fusion-based process is powder feedstock containing the pores which are formed during the gas atomization process [25, 26]. Both mechanical and fatigue properties were significantly affected by this type of porosity. The process–material–parameter effects on porosity is a highly complicated phenomenon, and a substantial amount of research work needs to be focused on that direction to control the porosity in fusion-based additive manufacturing processes [27–30].

Being a solid-state additive manufacturing process and since no melting or solidification is involved, the aforementioned porosities are generally absent in AFSD. However, incomplete material consolidation can potentially result in porosities in solid-state processing [15, 31]. The considerable material flow caused by the stirring action of the AFSD tool nevertheless results in effective interfacial bonding, meaning that the porosities are very

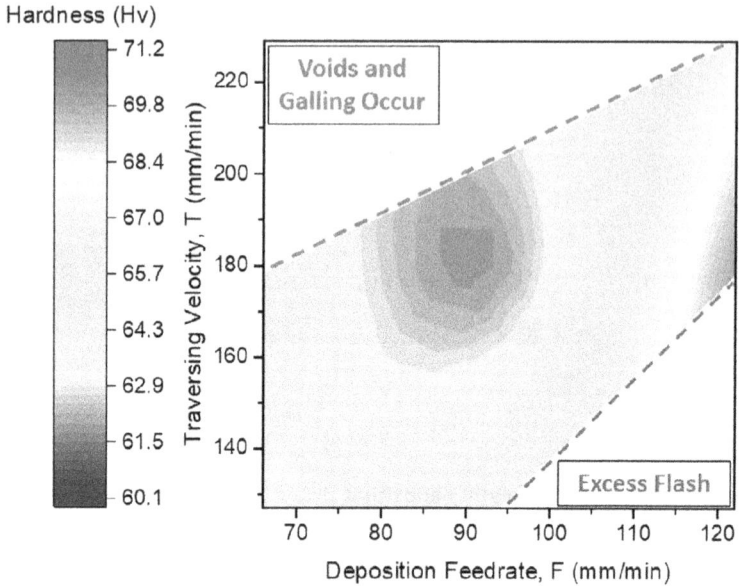

Figure 3.2 Plot showing the relationship between process parameters and defect formation [33].

much under control. Even if the feedstock is not fully dense, the printed component is free of porosity as a result of the severe plastic deformation and intense material flow under the AFSD tool. Farabi et al. reported a porosity of less than 0.01% in the AFSDed Ti6Al4V alloy revealing that a dense and homogeneous microstructure in the as-deposited state can be built using the AFSD process [32].

The void formation during AFSD is significantly influenced by process parameters. Insufficient material flow that prevents consolidation can cause voids and galling. This condition is caused by the combined effects of low material feed rate and high transverse velocity, which results in lower material deposition [33]. However, a higher material deposition rate can cause excessive flash formation. The relationship between process parameters and defects is better explained in Figure 3.2.

3.3.2 Microstructure

In the case of fusion-based AM techniques, a highly oriented, columnar grain structure results; by contrast, in AFSD, a well-refined, fine-grained microstructure is achieved because of the nature of thermomechanical processing. Figure 3.3 compares the resulting microstructure of Inconel625 alloy after the pulsed plasma arc deposition (PPAD) and the AFSD process. The PPAD process resulted in a columnar grain structure, whereas AFSD resulted in a consistent grain pattern. In AFSD, owing to the friction between the tool and material and also between the material and substrate, intense heat is generated

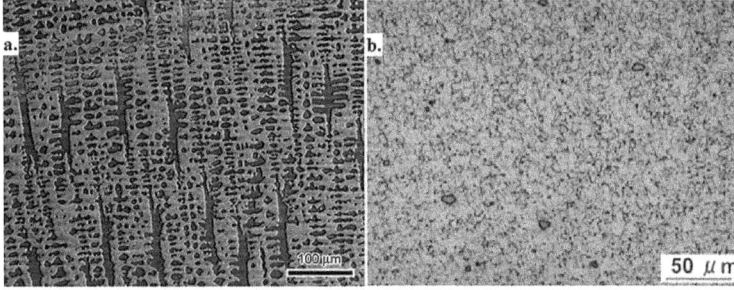

Figure 3.3 Microstructure of Inconel 625 alloy after (a) Pulsed plasma arc deposition (b) AFSD [34].

that can cause a temperature rise of about 0.6–0.9 times the melting point of the material to be deposited. This temperature rise is enough to cause intense plastic deformation and dynamic recrystallization which leads to an effective metallurgical bonding and well-refined, fine-grained microstructure.

In the case of aluminum and its alloys, continuous dynamic recrystallization is produced owing to the high-stacking fault energy, whereas for materials with low-stacking fault energy, discontinuous dynamic recrystallization is more significant. Both forms of the aforementioned dynamic recrystallization may lead to fine-grained equiaxed microstructure. However, microstructural inhomogeneity can develop during the AFSD process as the inner layer is subjected to reheating and re-deformation processes, unlike the surface layer, which does not undergo those processes.

A parametrization study was conducted by Willaiams et al. [35] to identify the effect of parameters on microstructural characterization in magnesium alloy WE43. The microstructure of WE43-T5 feedstock and as-deposited WE43 is compared in Figure 3.4. The average grain size of WE43 feedstock

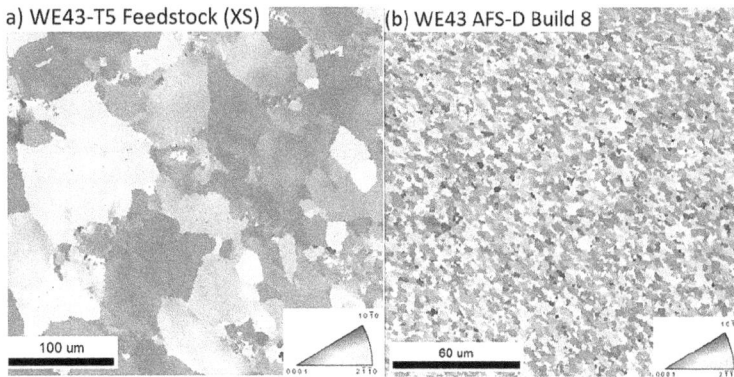

Figure 3.4 (a) EBSD image of WE43-T5 feedstock, (b) EBSD image of AFSD build WE43 [35].

is 45 microns whereas the same for the as-deposited WE43 is reduced to 2.7 microns exhibiting a 90% grain size reduction, owing to the dynamic recrystallization caused by the severe plastic deformation of the material. The WE43 feedstock exhibits an inhomogeneous grain morphology whereas AFSD build WE43 exhibits a more random texture and homogeneous grain morphology. A similar observation in the reduction of the grain size after the AFSD process on AZ31B magnesium alloy was reported by Joshi et al. [36]. The average grain size of the feed material is 13.5 microns whereas the same for the as-deposited material is 5 microns which is shown in Figure 3.5.

A similar trend was also exhibited in aluminum alloys during the AFSD process. Wedge et al. carried out the microstructural analysis of the AFSD process on AA2219 aluminum alloy and reported an average grain size reduction from 50 to 9 μm [37]. Similar observations of refinement in grain size were observed for the AFSD performed on various aluminum alloys [33, 38, 39]. In another research study, by Rivera et al., on the deposition of AA2219, grain refinement and the uniform distribution of grain size without

a. Feed material

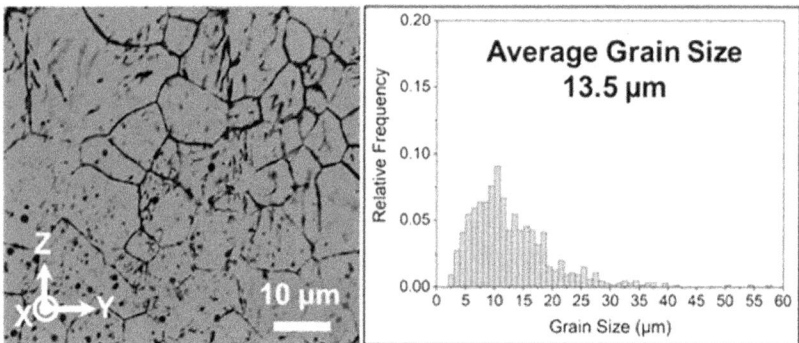

b. as-deposited

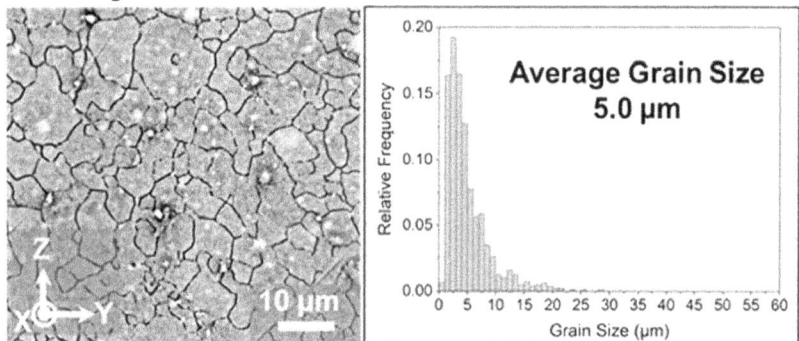

Figure 3.5 SEM micrographs and grain size distribution plot of AZ31B Magnesium alloy [36].

any appreciable variation between the layers was observed [38]. The average grain size at the top portion is 2.6 μm; the same measurements at the middle and bottom portions are 2.5 μm and 2.5 μm respectively.

Similar observations were also reported for titanium and Inconel alloys. Farabi et al. reported that significant grain refinement and homogeneous microstructure resulted while depositing Ti6Al4V at lower spindle speed using AFSD [32]. The average grain size of the as-received Inconel 625 is 30 μm, whereas after AFSD, the grain size reduces to 1 μm reporting a significant grain refinement [40].

Similar observations were also reported for titanium and Inconel alloys. Farabi et al. reported that significant grain refinement and homogeneous microstructure resulted while depositing Ti6Al4V at a lower spindle speed using AFSD [32]. The average grain size of the as-received Inconel 625 is 30 μm whereas after AFSD the grain size reduces to 1 μm reporting a significant grain refinement [40].

Though much of the research work gives a detailed analysis and comparison of the microstructures of the feed material and AFSD-built material, few investigations were also directed toward the microstructural evolution in transitional steps [41, 42]. The evolution of microstructure involves three main phases. First is the preheating phase, when the depositing material is inside the tool head. At this phase, the temperature rise happens because of the conduction phenomenon from the material substrate interface. During the deposition phase, the material, which is at an elevated temperature, is subjected to shear and compression and the elevated temperature reaches between 0.5 and 0.9 times the melting temperature. Severe plastic deformation occurs at this stage and conditions resembling hot working conditions cause dynamic recrystallization to dominate during this phase. Finally, during the third phase, a steady-state deposition is achieved and a deformation free period is reached as both the conduction and convection cool the material causing the post dynamic restoration to occur.

The investigation by Perry et al. revealed that grain refinement is more predominant during the initial material feeding where the grain size drastically reduces from 25–30 μm to 1–2 μm, whereas during the steady-state deposition no appreciable grain refinement is observed [41]. During the steady-state deposition, though strain accumulation is observed, further grain refinement is not reported because a dynamic equilibrium is attained between various competing processes.

The average grain size of the as-received and as-deposited materials after the AFSD process is tabulated in Table 3.1.

3.3.3 Mechanical properties

The severe plastic deformation and dynamic recrystallization resulted not only in grain refinement but also in enhanced mechanical properties when compared with other fusion-based AM processes. AFSD resulted in

Table 3.1 The grain size refinement of different materials after AFSD

Material	Average grain size (μm)	
	as-received	as-deposited
AZ31B Mg alloy [36]	13.5	5
Al-Mg-Si [42]	113	10
AA2024 [41]	30	2
WE43-T5 [35]	45	2.7
AA6061-T651 [33]	200	15
AA2219 [37]	50	9
IN625 [40]	30	1

an equiaxed grain structure which likely contributes to superior mechanical properties compared to other fusion-based AM processes. However, few studies revealed that micro hardness of the as-deposited materials are slightly less than that of the feed material which may be attributed to the dynamic recovery or precipitate evolution from the intrinsic high shear mechanism. Irregular thermal cycles also contribute to this slight decrease in micro hardness [35, 43, 44].

The hardness of the WE43 magnesium alloy deposition ranges from 75 to 85 HV at the different layers of deposition, as reported by Willaiams et al. [35], and is illustrated in Figure 3.6. From Figure 3.6, it is clear that a slightly higher hardness value was reported at the top two layers of the deposition. Other than this slight variation, there is a consistency in the hardness value throughout different layers of deposition. A slight decrease in hardness is

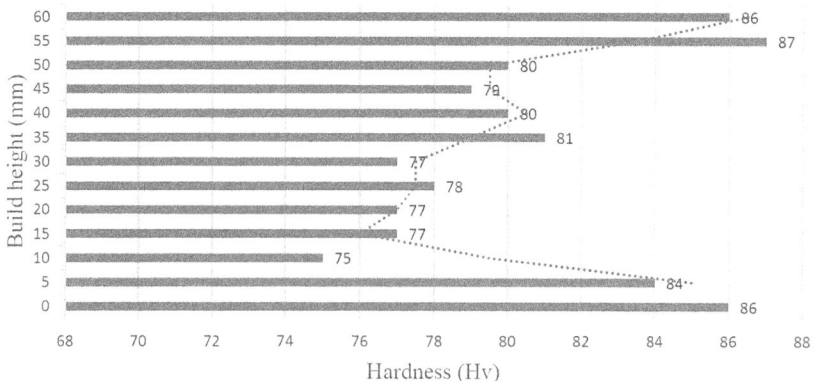

Figure 3.6 Hardness profile of WE43 magnesium alloy throughout the build profile.

observed in the middle section of deposition which can be noted from the Figure 3.6. This observation is due to the irregular thermal cycles.

A similar trend was also reported for an Al-Mg-Si alloy and a Cu alloy in the research work conducted by Griffiths et al. [42]. The main objective of the mentioned research work is to observe the uniformity of the mechanical properties after the AFSD process and reported that the standard deviation of the hardness value for Al-Mg-Si alloy is ± 1.47HV between the top and bottom layers of deposition and the same for Cu alloy is ± 1.17HV. Farabi et al. reported higher ductility and yield and ultimate tensile strengths of Ti6Al4V alloy processed via AFSD than that manufactured by fusion-based AM process [32]. Joshi et al. reported a marginal increase in hardness of the as-deposited AZ31B Mg alloy (58 ± 1.7 HV) when compared to the feed material (53 ± 2 HV). This marginal increase in hardness can be attributed to grain refinement.

3.3.4 Residual stresses

One of the inherent limitations associated with the fusion-based AM processes is the anisotropic residual stresses developed in the deposited components. Large thermal gradients in the fusion-based AM processes because of the melting and solidification involved resulted in high residual stresses in the as-deposited components. This significantly affects the mechanical properties of the printed components, particularly the impact fatigue properties, and can also cause geometric distortion. To alleviate this issue, post-process heat treatment is required to relieve the residual stresses. However, this adds cost to an already expensive process.

Since AFSD is a solid-state processing that does not involve melting and solidification, residual stresses are almost absent. Owing to this, not just the residual stresses but also the solidification-related defects, including hot cracking, aggregation of oxide particles, etc., are absent. Hence heat-treatment processes are not required after AFSD. However, since AFSD is a near-net shaping process, post-processing is required for obtaining a high surface finish.

3.3.5 Feed material

Feed material in the powder form is used as an input in most of the fusion-based AM processes. A substantial amount of literature reported that the final quality of manufactured product using the fusion-based AM process is strongly influenced by the powder quality attributable to the shape and size distribution of the individual particulates and chemical composition [45–48]. This expensive and time-consuming powder preparation adds further cost to the process.

With AFSD, feed material need not necessarily be in powder form; even a solid rod can be fed through the rotating tool for deposition. Any solid bar, cast or wrought alloys, can be fed as input in AFSD; thereby, the effort and cost spent on pre-processing can be eliminated. Even if the feed material in powder form is used as input to AFSD, there is no need to adhere to the stringent requirements of powder attributes as in the case with fusion-based AM processes. Powder feed material containing pores is not of concern as AFSD is a solid-state thermomechanical process that yields a deposited structure with no porosity, residual stresses, and other solidification-related defects.

3.3.6 Energy consumption, cost, and build rate

The high-energy laser or electron beam sources used in fusion-based additive manufacturing processes to melt the feed material lead to a high-energy pathway for additive manufacturing. Such high-energy sources are not required in AFSD and it relies on simple mechanical principles and resembles more like a CNC milling machine.

The build rate of AFSD is several times higher than most of the other fusion-based additive manufacturing processes in the current market [49]. A comparison of the build rate of AFSD with other metal additive manufacturing techniques for different materials is presented in Table 3.2. Scalability is another factor in favor of the AFSD process. In other fusion-based additive maintaining the build chamber atmosphere needs to be carefully dealt with as there is a tendency for metal powders to oxidize when they come into contact with air. Inert gas/vacuum is required in the build chamber, which naturally limits the scalability of the fusion-based additive manufacturing processes. However, AFSD is an open-air additive manufacturing process and does not require complicated and expensive maintenance of the build chamber atmosphere. Without these constraints, large-scale components on meter scale can be deposited using this technique. The Meld Manufacturing Corporation recently printed a 10.5ft diameter aluminum cylinder using the AFSD process that demonstrates the scalability of the process. Some of the large components that are fabricated using AFSD are shown in Figure 3.7.

Table 3.2 Build rate of different AM techniques for different materials [50]

	Build rate (cm³/hr)			
	Aluminum	Steel	Nickel	Titanium
AFSD	1020	622	81.8	553
LPBS	2.52	5	0.18	1.26
LBD	4.06		5.54	4.83
EBM	80	14.3		15.63

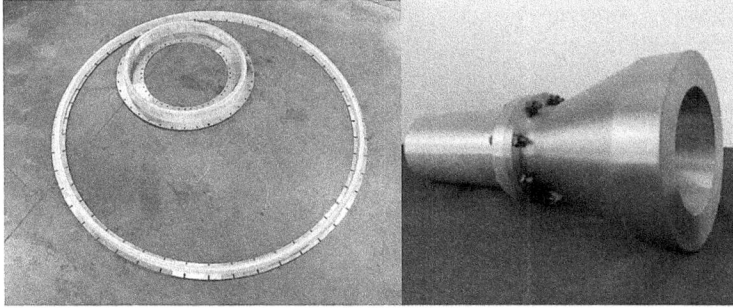

Figure 3.7 Large components fabricated using AFSD process [51].

The time-consuming and expensive powder preparation is not required in printing using AFSD. In addition to that, there is no need to maintain a build chamber atmosphere or to carry out post-processing heat treatment. For these reasons, the cost associated with printing using AFSD is relatively low. However, AFSD is a near-net shaping process, and additional finishing operations are required to achieve superior surface quality.

3.3.7 Wide range of materials and applications

Fusion-based additive manufacturing techniques for non-weldable alloys are still in the development stages and tasted limited success for now. Solidification involved and the related defects, particularly the hot-cracking problem, restrict its applicability to aluminum alloys. However, such restrictions are not applicable to AFSD. Not just the aluminum alloys but the high-strength materials such as nickel and titanium-based alloys can also be fabricated using AFSD [52]. Some of the few materials that are successfully fabricated using AFSD include Aluminum (1XXX, 2XXX, 5XXX, 6XXX, 7XXX), Copper (Cu110, NAB), Steel (ODS 14YWT-F82H, 300M-4140, Aermet 100-4140, A514), Nickel (Inconel 625, Inconel 718, Inconel 600, Cu-Ni 200, Nimonic 80A), Magnesium (AZ31, AZ91, WE43, E675, AMX602, E21), Titanium (Ti64, Titanium Aluminide) MMC's (Al-SiC, Al-Fe, Al-W, Al-Mo, Al-BNC, Al-CF, Al-CNT, Cu-W, Cu-Ta), etc.

Since AFSD resulted in a refined, equiaxed microstructure with enhanced mechanical properties, the AFSD can target the applications where enhanced mechanical properties with the least post-processing heat treatment are preferred. Good scalability makes AFSD particularly suitable for the structural repair of large components, particularly in the aerospace and defense sector. Griffiths et al., through their research work on the repair of volume damages in AA7075, successfully demonstrated the feasibility of AFSD in large-scale repair works [53]. This research work reported promising results in filling the entire volume of through holes and wide grooves in AA7075 alloy despite some imperfect repair quality. AFSD also enables sufficient mixing during the repair work, which is evident by the fine equiaxed grain structure. This was one of the preliminary investigations undertaken on the repair work of AFSD and hence no optimization strategy was applied for the process parameters and scanning path, resulting in imperfect repair quality in lower portions compared to upper portions. Also, a less than 15% decrease in the hardness of the repaired volume is reported in comparison with the original feed material. Further works conducted by the same group of authors reported that such defects were largely eliminated by optimizing the process parameters (rotation speed of the tool and dwell time). By adopting systematic optimization strategies on process parameters, high-quality and highly flexible repair works can be performed using AFSD.

AFSD can also be used for cladding purposes without any melting involved and thus can achieve all the positives of solid-state processing. AFSD can ensure a strong metallurgical bond between the materials. Hartley et al. demonstrated the feasibility of applying AFSD for cladding on thin Al-Mg-Si sheet metals [54]. A single-layer cladding 0.9 mm thick is successfully produced on a 1.4 mm-thick sheet without any buckling or wrinkling. Good mixing between the cladding and the substrate, uniform grain structure, and no residual stresses are all reported in this work, thanks to the solid-state processing. Figure 3.8 shows an example of AFSD applied for

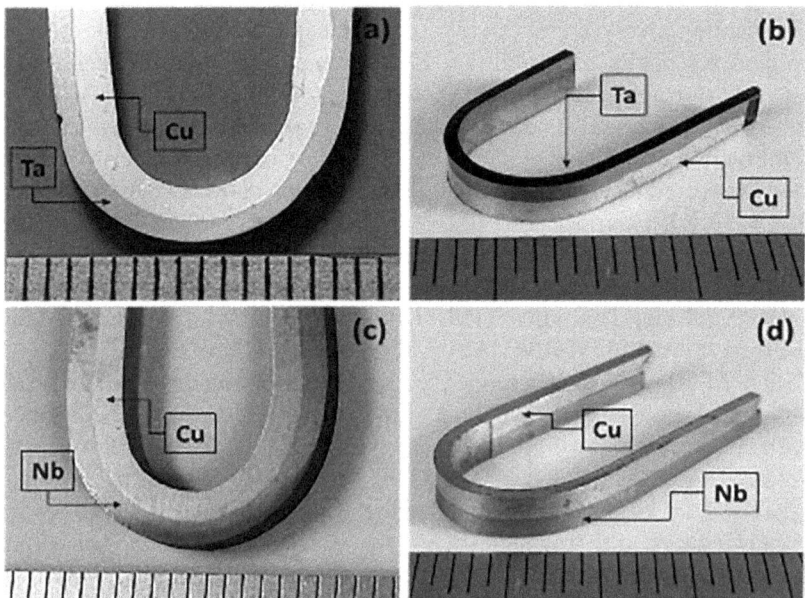

Figure 3.8 Tantalum and niobium coated with copper [55].

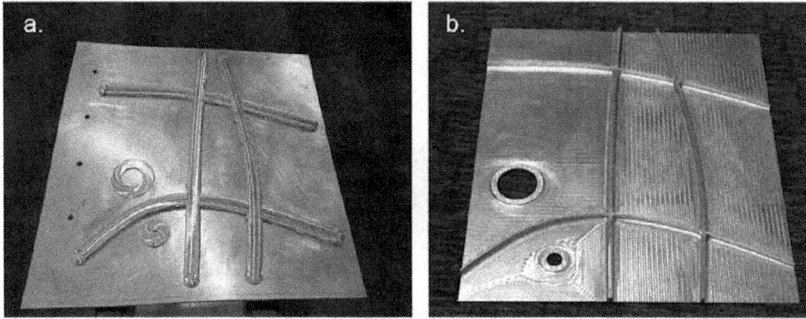

Figure 3.9 Rib stiffeners added using AFSD (a) as-deposited panel, (b) final machined panel [10].

cladding operation. The application area of AFSD can also be extended to the addition of features including flanges, rib stiffeners, attachment points, etc., which it is difficult to incorporate into casting or extrusion operations. An example of adding rib stiffeners using AFSD is shown in Figure 3.9. This allows design engineers to use thinner plate materials for lightweight components and strengthening features can be later incorporated at strategic locations using AFSD.

3.4 CURRENT LIMITATIONS

Despite the great benefits offered, AFSD is still in the elementary stages and many more fundamental aspects of the process need to be completely revealed. Currently, there are several limitations associated with this process which are explained below.

3.4.1 Resolution

One of the main aspects where AFSD falls short compared to fusion-based additive manufacturing processes is the in-plane resolution. The in-plane resolution of many of the fusion-based processes is 0.5 mm whereas the same for AFSD, with the current tooling capability, is in the range of 10 mm. Reducing the shoulder size of the tool, and thereby improving the resolution, cannot be considered a feasible option as this may adversely affect the material consolidation, material flow, and build rate. Contact friction between the tool and the substrate largely determines the heat generation and a smaller contact area leads to less heat generation and thereby poor material flow and material consolidation. Therefore, AFSD can be considered as a metal additive manufacturing solution for large-size components with a high build rate and superior properties but at the sacrifice of resolution.

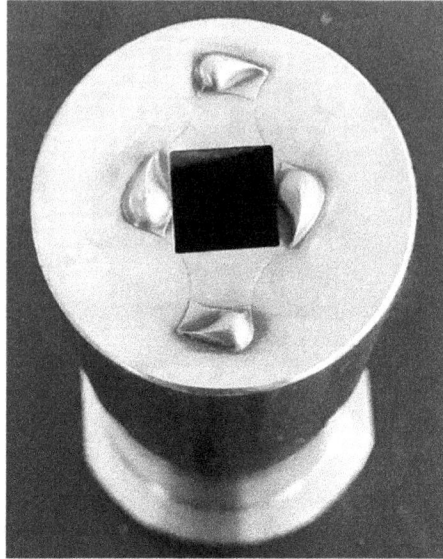

Figure 3.10 AFSD tool with teardrop protrusion [37].

3.4.2 Tooling

AFSD shares many principles of FSW/P and the tooling configuration is widely investigated in FSW/P. Only a few configurations were so far researched for AFSD and the most common is the flat bottom face of the hollow shoulder tool. However, recent research works reported that small protrusions provided on the bottom face of the tool can enhance material flow and mixing. AFSD tool with teardrop protrusion is shown in Figure 3.10. A deeper understanding of the various tool configurations and their influence on the material flow and grain refinement are the future prospects for research.

Tool steels are commonly used as the tool material when processing low-strength materials whereas tungsten carbide or Polycrystalline Cubic Boron Nitride (PCBN) are commonly adopted when processing high-strength materials. Tool shoulders made of, or coated with, advanced ceramics or composites have to be developed to process materials having high strength and melting temperatures. This turns out to be expensive but a trade-off must be made between the economic viability and the wear rate of the shoulder. This necessitates the need for further research work not in the tooling configurations but also in the tooling materials.

An AFSD tool is subjected to extreme conditions and hence wear will occur, but none of the research studies were reported on quantifying the tool wear during the AFSD process. As tooling is very significant in AFSD and the efficiency of the deposition largely depends on the tooling capability, tool wear needs to be extensively studied which opens another area of research.

3.4.3 Near net-shaping

AFSD is a near-net-shaping process and the resolution is in terms of centime-ter scale. Large components with complex 3D geometries can be fabricated by using AFSD but the surface finish will be poor. If an improved surface finish is desired, then post-processing machining and finishing operations are required. Developing a hybrid process combining AFSD and CNC mill-ing can make this process move closer to the net shaping process with better resolution and without sacrificing the build rate.

3.5 PATHWAYS TO FUTURE IMPROVEMENTS IN AFSD

Despite some of the promising characteristics of metal additive manufactur-ing, AFSD is yet to make an impact on commercial applications. There are perhaps two main reasons for this: One is that this is a comparatively new technology and many are unaware as yet of its unique characteristics. The second is that AFSD is still in the pre-mature stage and the basic underly-ing principles are yet to be completely understood. Lot of opportunities for R&D are there to transform AFSD into a viable alternative for fusion-based AM processes. Some of the future research areas are proposed below.

3.5.1 Processing parameters

Optimizing the processing parameters is very crucial in any fabrication pro-cess. AFSD is still in the initial stages and a clear understanding of the influ-ence of significant parameters on the final quality of the deposition remains unclear. The key processing parameters that influence the quality of deposi-tion in AFSD are tool rotation speed (ω), feedstock feed rate (F), and tool traversing velocity (V). Different build temperatures can be attained using different parametric combinations. Though the peak temperature attained during the process is less than the melting temperature, it can cause the dis-solution of the primary strengthening precipitates. Therefore, optimizing the AFSD parameters is very crucial in achieving quality deposition with desired mechanical properties. In order to build components using AFSD with tai-lor-made microstructure and mechanical properties, a complete understand-ing of how each parameter is influencing the deposition is needed.

Generally, the heat generation rate and strain rate are more controlled by the ω and F, whereas V determines the time available for deformation. F also has a direct influence on the deposition rate. The general process window for AFSD is shown in Figure 3.11. Large ω can lead to a damaging of the tool head, the jamming of feed material inside the tool shoulder, and also, in extreme cases, it can even lead to melting. However, lower ω can cause inadequate heat genera-tion and thereby may result in inadequate plastic deformation. Low F and high V can lead to the deposition with poor surface quality and more pores.

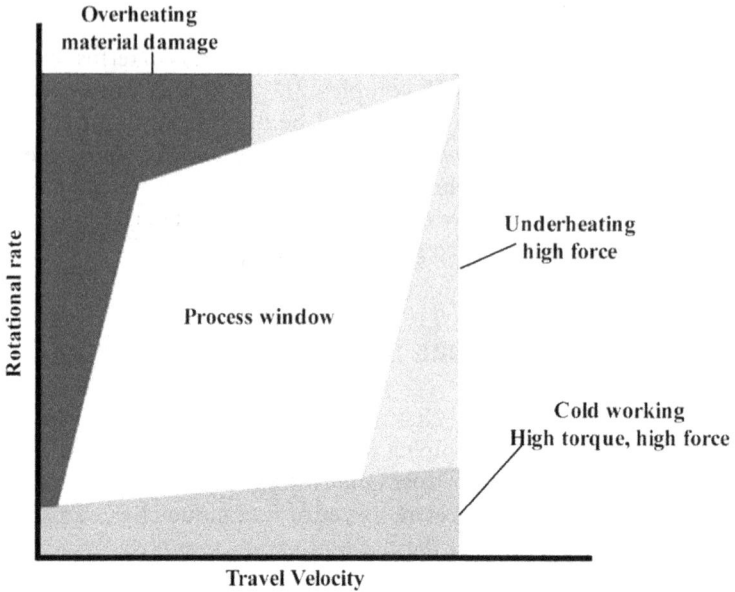

Figure 3.11 Feasible process window for AFSD [42].

Stubblefield et al. reported that the feedstock feed rate was more influential in affecting the quality of final deposition [56]. The attained temperature is directly proportional to the federate with a higher temperature being the result of a higher federate. Inadequate material flow caused by lower feedstock feed rate resulted in less frictional heat generated between the contact surface of the tool and the substrate and thereby resulted in lower temperature attained during the deposition. Higher feedstock federates also cause higher downward force, generating more frictional heat. Feed rate also influences the homogeneous deposition, with lower federate would likely result in a heterogeneous deposition because of insufficient material mixing. Increasing the feed rate can lead to a homogeneous deposition, however, increasing the feed rate beyond an optimal point may lead to material wastage through excessive flash. Willaiams et al. reported that a higher deposition ratio (V/F) for magnesium alloy WE43 resulted in a favorable deposition [35].

Griffiths et al. studied the process–microstructure linkages on Al-Mg-Si alloy and Cu [42]. In the case of the Al-Mg-Si alloy, a higher ω/V ratio gives a higher strain level and thereby increases the degree of recrystallization. However, the strain level for Cu is more dependent on V, with a lower strain level for higher values of V and in such a condition microstructure evolution is more dominated by post-dynamic recrystallization. Farabi et al. reported that for the deposition of Ti6Al4V alloy, ω chiefly influences the peak temperature attained and lower ω leads to significant grain refinement.

Only a few works were reported so far for linking the process parameters with microstructure evolution and mechanical properties. Plenty of research opportunities are ahead in establishing such linkages and thereby obtaining a clear understanding of the physics underlying the process.

3.5.2 Hybrid process

Being a net-shaping process, the surface quality is poor during the deposition in AFSD, and so post-process finishing operation is required. The resolution in the AFSD process is also very low, which is attributed to the large size of the tool shoulder. However, a reduction in tool size to improve the resolution is not feasible as it may impact the build rate. One innovative solution to overcome both the above-mentioned limitations is to develop a hybrid process combining AFSD and conventional machining operations, especially milling. Currently, hybrid manufacturing systems with a laser-based metal printing unit that can be integrated with industrial motion hardware such as CNC machines and robotic cells are available in the market [57]. Since the AFSD machine partially resembles and shares common principles with the CNC milling machine, combining these two techniques can open up a new area of opportunities. This could help in improving the resolution as well as bringing AFSD closer to a net-shaping process.

3.5.3 In-process monitoring

Substantial efforts have to be made for developing in-process monitoring techniques to improve the quality and reliability of the deposition. Few attempts are currently underway for in-process monitoring by integrating multiple sensors, including infrared cameras for monitoring thermal evolution, 3D laser scanners for monitoring the surface topography, ultrasonic testing to monitor the defect levels, etc. Closed-loop control can also be implemented to automatically modify the parameters in mitigating any anomaly. Furthermore, online monitoring and controlling of the temperature attained and the cooling rate is one of the most challenging research topics and requires much further attention. This will enable the widespread adoption of AFSD for metal additive manufacturing solutions.

3.5.4 In-depth understanding of basic principles

Since AFSD is a solid-state thermomechanical process, thermal evolution and plastic deformation are vital in determining the quality of deposition. The two aspects of AFSD which need to be extensively researched are the thermal deformation history (variation of temperature and strain as a function of time) and the influence of process parameters on the evolution of temperature and plastic deformation. Infrared imaging can be employed for capturing the thermal evolution. The results of preliminary research in this aspect were reported by Garcia et al. in which the peak temperature, reheating and cooling rate were monitored during the deposition and established an empirical relation linking the process parameters and peak temperature [44]. Monitoring the thermal evolution, however, is a straightforward technique whereas monitoring the deformation history is more challenging as the plastic deformation is a rapid phenomenon. To monitor the deformation and material flow in the internal surface is more complicated and a feasible solution to this is to place a tracer material in the feedstock followed by the real time characterization of tracer shape evolution. Many more research works have to be initiated in this direction to get a better understanding of the general deformation characteristics of AFSD process.

3.5.5 Tooling

The present tooling capability of AFSD is explained under Section 3.4.2. As the tool configuration can directly influence the heat generation and material mixing, some detailed study on the different configurations and designs of the tools is required. Tooling materials made of advanced ceramics or composites can also be used for processing high-strength materials. This is particularly a key phenomenon when applying AFSD as a composite fabrication technique.

3.6 SUMMARY

AFSD is a solid-state thermomechanical metal additive manufacturing process that has the potential to troubleshoot some of the inherent limitations associated with fusion-based additive manufacturing processes. AFSD is particularly suited to printing of large-scale complex 3D geometries with fine, equiaxed microstructure and enhanced mechanical properties. Both weldable and non-weldable alloys can be deposited using AFSD. The main application areas of AFSD is not just limited to fabrication 3D structures but can also be extended to structural repair, adding features and cladding/coating operations. There are two factors presently in favor of AFSD: (i) the substantial increase in build rate compared to fusion-based metal additive manufacturing techniques and (ii) the good scalability. Current areas in need of improvement include poor surface quality and lower in-plane resolution. Despite the unique characteristics of AFSD, the potential of the process is yet to be fully exploited, meaning that there are tremendous opportunities for R&D. In-depth understanding of the underlying basic principles of AFSD, establishing the process parameter–microstructure–properties linkage, in-situ monitoring and improving the tooling materials and capabilities are among key research areas that require immediate attention. Developing a hybrid process combining AFSD and CNC milling can help overcome some of the limitations of AFSD.

Considering the unique capabilities of AFSD, it can be easily argued that the process has the full potential to develop into a metal additive manufacturing solution for aerospace, defense and automotive applications. However, the AFSD process is still in its pre-mature stage and numerous R&D works have to be undertaken to ensure the widespread implementation of this process for commercial applications.

REFERENCES

[1] R. Rai, M. K. Tiwari, D. Ivanov and A. Dolgui, "Machine learning in manufacturing and industry 4.0 applications," *International Journal of Production Research*, vol. 59, no. 16, pp. 4773–4778, 2021.

[2] M. Sanchez, E. Exposito and J. Aguilar, "Industry 4.0: survey from a system integration perspective," *International Journal of Computer Integrated Manufacturing*, vol. 33, no. 10–11, pp. 1017–1041, 2020.

[3] L. D. Xu, E. L. Xu and L. Li, "Industry 4.0: state of the art and future trends," *International Journal of Production Research*, vol. 56, no. 8, pp. 2941–2962, 2018.

[4] H. Fatorachian and H. Kazemi, "Impact of Industry 4.0 on supply chain performance," *Production Planning & Control*, vol. 32, no. 1, pp. 63–81, 2021.

[5] H. Parmar, T. Khan, F. Tucci, R. Umer and P. Carlone, "Advanced robotics and additive manufacturing of composites: towards a new era in Industry 4.0," *Materials and Manufacturing Processes*, vol. 37, no. 5, pp. 483–517, 2022.

[6] "Additive manufacturing—General principles—Fundamentals and vocabulary," ISO/ASTM 52900:2021, 11 2021. [Online]. Available: https://www.iso.org/standard/74514.html

[7] Z. Chen, C. Han, M. Gao, S. Y. Kandukuri and K. Zhou, "A review on qualification and certification for metal additive manufacturing," *Virtual and Physical Prototyping*, vol. 17, no. 2, pp. 382–405, 2022.

[8] Y. Zhang, S. Yang and Y. F. Zhao, "Manufacturability analysis of metal laser-based powder bed fusion additive manufacturing—a survey," *The International Journal of Advanced Manufacturing Technology*, vol. 110, pp. 57–78, 2020.

[9] D. Zhang, A. Prasad, M. J. Bermingham, C. J. Todaro, M. J. Benoit, M. N. Patel, D. Qiu, D. H. St. John, M. Qian and M. A. Easton, "Grain refinement of alloys in fusion-based additive manufacturing processes," *Metallurgical and Materials Transactions A*, vol. 51, pp. 4341–4359, 2020.

[10] V. Gopan, K. L. D. Wins and A. Surendran, "Innovative potential of additive friction stir deposition among current laser based metal additive manufacturing processes: a review," *CIRP Journal of Manufacturing Science and Technology*, vol. 32, pp. 228–248, 2021.

[11] W. J. Sames, F. A. List, S. Pannala, R. R. Dehoff and S. S. Babu, "The metallurgy and processing science of metal additive manufacturing," *International Materials Reviews*, vol. 61, no. 5, pp. 315–360, 2016.

[12] M. Srivastava, S. Rathee, S. Maheshwari, A. N. Siddiquee and T. K. Kundra, "A review on recent progress in solid-state friction based metal additive manufacturing: Friction stir additive techniques," *Critical Reviews in Solid-State and Materials Sciences*, vol. 44, no. 5, pp. 345–377, 2019.

[13] H. Lee, C. H. J. Lim, M. J. Low, N. Tham, V. M. Murukeshan and Y.-J. Kim, "Lasers in additive manufacturing: a review," *International Journal of Precision Engineering and Manufacturing-Green Technology*, vol. 4, pp. 307–322, 2017.

[14] J. Gardan, "Additive manufacturing technologies: state of the art and trends," *International Journal of Production Research*, vol. 54, no. 10, pp. 3118–3132, 2016.

[15] R. S. Mishra, R. S. Haridas and P. Agrawal, "Friction stir-based additive manufacturing," *Science and Technology of Welding and Joining*, vol. 27, no. 3, pp. 141–165, 2022.

[16] W. Li, C. Cao, G. Wang, F. Wang, Y. Xu and X. Yang, "'Cold spray +' as a new hybrid additive manufacturing technology: a literature review," *Science and Technology of Welding and Joining*, vol. 24, no. 5, pp. 420–445, 2019.

[17] J. Obielodan and B. Stucker, "A fabrication methodology for dual-material engineering structures using ultrasonic additive manufacturing," *The International Journal of Advanced Manufacturing Technology*, vol. 70, pp. 277–284, 2014.

[18] J. B. Jordon, H. Rao, R. Amaro and P. Allison, "Beyond friction stir welding: friction stir processing and additive manufacturing," *Fatigue in Friction Stir Welding*, pp. 137–143, 2019.

[19] A. Heidarzadeh, S. Mironov, R. Kaibyshev, G. Çam, A. Simar, A. Gerlich, F. Khodabakhshi, A. Mostafaei, D. P. Field, D. Robson, A. Deschamps and P. J. Withers, "Friction stir welding/processing of metals and alloys: A comprehensive review on microstructural evolution," *Progress in Materials Science*, vol. 117, p. 100752, 2021.

[20] S. Rathee, M. Srivastava, P. M. Pandey, A. Mahawar and S. Shukla, "Metal additive manufacturing using friction stir engineering: a review on microstructural evolution, tooling and design strategies," *CIRP Journal of Manufacturing Science and Technology*, vol. 35, pp. 560–588, 2021.

[21] "MELD Manufacturing," [Online]. Available: https://www.meldmanufacturing.com/

[22] J. Kunz, S. Herzog, A. Kaletsch and C. Broeckmann, "Influence of initial defect density on mechanical properties of AISI H13 hot-work tool steel produced by laser powder bed fusion and hot isostatic pressing," *Powder Metallurgy*, vol. 65, no. 1, pp. 1–12, 2022.

[23] M. Montazeri, A. R. Nassar, A. J. Dunbar and P. Rao, "In-process monitoring of porosity in additive manufacturing using optical emission spectroscopy," *IISE Transactions*, vol. 52, no. 5, pp. 500–515, 2020.

[24] N. H. Paulson, B. Gould, S. J. Wolff, M. Stan and A. C. Greco, "Correlations between thermal history and keyhole porosity in laser powder bed fusion," *Additive Manufacturing*, vol. 34, p. 101213, 2020.

[25] S. Shresth and K. Chou, "Formation of keyhole and lack of fusion pores during the laser powder bed fusion process," *Manufacturing Letters*, vol. 32, pp. 19–23, 2022.

[26] W. Hearn, R. Steinlechner and E. Hryha, "Laser-based powder bed fusion of non-weldable low-alloy steels," *Powder Metallurgy*, vol. 65, no. 2, pp. 121–132, 2022.

[27] M. Giovagnoli, G. Silvi, M. Merlin and M. T. D. Giovanni, "Optimisation of process parameters for an additively manufactured AlSi10Mg alloy: Limitations of the energy density-based approach on porosity and mechanical properties estimation," *Materials Science and Engineering: A*, vol. 802, p. 140613, 2021.

[28] L. Zhao, L. Song, J. G. S. Macías, Y. Zhu, M. Huang, A. Simar and Z. Li, "Review on the correlation between microstructure and mechanical performance for laser powder bed fusion AlSi10Mg," *Additive Manufacturing*, vol. 56, p. 102914, 2022.

[29] W. H. Kan, L. N. S. Chiu, C. V. S. Lim, Y. Zhu, Y. Tian, D. Jiang and A. Huang, "A critical review on the effects of process-induced porosity on the mechanical properties of alloys fabricated by laser powder bed fusion," *Journal of Materials Science*, vol. 57, pp. 9818–9865, 2022.

[30] J. P. Oliveira, A. D. L. Londe and J. Ma, "Processing parameters in laser powder bed fusion metal additive manufacturing," *Materials & Design*, vol. 193, p. 108762, 2020.

[31] F. Khodabakhshi and A. P. Gerlich, "Potentials and strategies of solid-state additive friction-stir manufacturing technology: a critical review," *Journal of Manufacturing Processes*, vol. 36, pp. 77–92, 2018.

[32] E. Farabi, S. Babaniaris, M. R. Barnett and D. M. Fabijanic, "Microstructure and mechanical properties of Ti6Al4V alloys fabricated by additive friction stir deposition," *Additive Manufacturing Letters*, vol. 2, p. 100034, 2022.

[33] B. J. Phillips, D. Z. Avery, T. Liu, O. L. Rodriguez, C. J. T. Mason, J. B. Jordon, L. N. Brewer and P. G. Allison, "Microstructure-deformation relationship of additive friction stir-deposition Al–Mg–Si," *Materialia*, vol. 7, p. 100387, 2019.

[34] F. Xu, Y. L. Liu, F. Shu, P. He and B. Xu, "Microstructural evolution and mechanical properties of Inconel 625 alloy during pulsed plasma arc deposition process," *Journal of Materials Science & Technology*, vol. 29, no. 5, pp. 480–488, 2013.

[35] M. B. Willaiams, T. W. Robinson, C. Williamson, R. Kinser, N. A. Ashmore, P. G. Allison and J. B. Jordon, "Elucidating the effect of additive friction stir deposition on the resulting microstructure and mechanical properties of magnesium alloy WE43," *Metals*, vol. 11, p. 1739, 2021.

[36] S. S. Joshi, S. M. Patil, S. Mazumder, S. Sharma, D. A. Riley, S. Dowden, R. Banerjee and N. B. Dahotre, "Additive friction stir deposition of AZ31B magnesium alloy," *Journal of Magnesium and Alloys*, vol. 10, 2022.

[37] K. A. Wedge, D. Z. Avery, S. R. Daniewicz, J. W. Sowards, P. G. Allison, J. B. Jordon and R. L. Amaro, "Characterization of the fatigue behavior of additive friction stir-deposition AA2219," *International Journal of Fatigue*, vol. 142, p. 105951, 2021.

[38] O. G. Rivera, P. G. Allison, L. N. Brewer, O. L. Rodriguez, J. B. Jordon, T. Liu, W. Whittington, R. L. Martens, Z. McClelland, C. J. T. Mason, L. Garcia, J. Q. Sug and N. Hardwick, "Influence of texture and grain refinement on the mechanical behavior of AA2219 fabricated by high shear solid-state material deposition," *Materials Science and Engineering: A*, vol. 724, pp. 547–558, 2018.

[39] M. E. J. Perry, R. J. Griffiths, D. Garcia, J. M. Sietins, Y. Zhu and H. Z. Yu, "Morphological and microstructural investigation of the non-planar interface formed in solid-state metal additive manufacturing by additive friction stir deposition," *Additive Manufacturing*, vol. 35, p. 101293, 2020.

[40] O. G. Rivera, P. G. Allison, J. B. Jordon, O. L. Rodriguez, L. N. Brewer, Z. McClelland, W. R. Whittington, D. Francis, J. Su, R. L. Martens and N. Hardwick, "Microstructures and mechanical behavior of Inconel 625 fabricated by solid-state additive manufacturing," *Materials Science and Engineering: A*, vol. 694, pp. 1–9, 2017.

[41] M. E. J. Perry, H. A. Rauch, R. J. Griffiths, D. Garcia, J. M. Sietins, Y. Zhu, Y. Zhu and H. Z. Yu, "Tracing plastic deformation path and concurrent grain refinement during additive friction stir deposition," *Materialia*, vol. 18, p. 101159, 2021.

[42] R. J. Griffiths, D. Garcia, J. Song, V. K. Vasudevan, M. A. Steiner, W. Cai and H. Z. Yu, "Solid-state additive manufacturing of aluminum and copper using additive friction stir deposition: process-microstructure linkages," *Materialia*, vol. 15, p. 100967, 2021.

[43] D. Z. Avery, O. G. Rivera, C. J. T. Mason, B. J. Phillips, J. B. Jordon, J. Su, N. Hardwick and P. G. Allison, "Fatigue behavior of solid-state additive manufactured Inconel 625," *JOM*, vol. 70, pp. 2475–2484, 2018.

[44] D. Garcia, W. D. Hartley, H. A. Rauch, R. J. Griffiths, R. Wang, Z. J. Kong, Y. Zhu and H. Z. Yu, "In situ investigation into temperature evolution and heat generation during additive friction stir deposition: A comparative study of Cu and Al-Mg-Si," *Additive Manufacturing*, vol. 34, p. 101386, 2020.

[45] A. C. Field, L. N. Carter, N. J. E. Adkins, M. M. Attallah, M. J. Gorley and M. Strangwood, "The effect of powder characteristics on build quality of high-purity tungsten produced via laser powder bed fusion (LPBF)," *Metallurgical and Materials Transactions A*, vol. 51, p. 1367–1378, 2020.

[46] M. Lebba, A. Astarita, D. Mistretta, I. Colonna, M. Liberini, F. Scherillo, C. Pirozzi, R. Borrelli, S. Franchitti and A. Squillace, "Influence of powder characteristics on formation of porosity in additive manufacturing of Ti-6Al-4V components," *Journal of Materials Engineering and Performance*, vol. 26, pp. 4138–4147, 2017.

[47] N. Pirch, S. Linnenbrink, A. Gasser and H. Schleifenbaum, "Laser-aided directed energy deposition of metal powder along edges," *International Journal of Heat and Mass Transfer*, vol. 143, p. 118464, 2019.

[48] O. D. Neikov, "Atomization and Granulation," in *Handbook of Non-Ferrous Metal Powders (Second Edition) Technologies and Applications*, 2019, pp. 125–185.

[49] H. Z. Yu, M. E. Jones, G. W. Brady, R. J. Griffiths, D. Garcia, H. A. Rauch, C. D. Cox and N. Hardwick, "Non-beam-based metal additive manufacturing enabled by additive friction stir deposition," *Scripta Materialia*, vol. 153, pp. 122–130, 2018.

[50] "Meld Manufacturing," [Online]. Available: http://meldmanufacturing.com/wp-content/uploads/2018/08/MELD-Overview.pdf

[51] "Meld Manufacturing (retrieved on November 29, 2022)," [Online]. Available: https://www.meldmanufacturing.com/

[52] R. J. Griffiths, M. E. J. Perry, J. M. Sietins, Y. Zhu, N. Hardwick, C. D. Cox, H. A. Rauch and H. Z. Yu, "A perspective on solid-state additive manufacturing of aluminum matrix composites using MELD," *Journal of Materials Engineering and Performance*, vol. 28, pp. 648–656, 2018.

[53] R. J. Griffiths, D. T. Petersen, D. Garcia and H. Z. Yu, "Additive friction stir-enabled solid-state additive manufacturing for the repair of 7075 aluminum alloy," *Applied Sciences*, vol. 9, no. 17, p. 3486, 2019.

[54] W. D. Hartley, D. Garcia, J. K. Yoder, E. Poczatek, J. H. Forsmark, S. G. Luckey, D. A. Dillard and H. Z. Yu, "Solid-state cladding on thin automotive sheet metals enabled by additive friction stir deposition," *Journal of Materials Processing Technology*, vol. 291, p. 117045, 2021.

[55] "Meld Manufacturing (retrieved on November 29, 2022)," [Online]. Available: https://www.meldmanufacturing.com/coat/

[56] G. G. Stubblefield, K. Fraser, B. J. Phillips, J. B. Jordon and P. G. Allison, "A meshfree computational framework for the numerical simulation of the solid-state additive manufacturing process, additive friction stir-deposition (AFS-D)," *Materials & Design*, vol. 202, p. 109514, 2021.

[57] "3D Monotech, Meltio Engine CNC Integration," [Online]. Available: https://3dmonotech.in/meltio-engine-cnc-integration/

Chapter 4

Cold spray additive manufacturing (CSAM)

Wen Sun

Hubei Chaozhuo Aviation Technology Co., Ltd., Xiangyang, China

Xin Chu

Institute of New Materials, Guangdong Academy of Sciences, Guangzhou, China

Haiming Lan

Hubei Chaozhuo Aviation Technology Co., Ltd., Xiangyang, China

Jibo Huang

South China University of Technology, Guangzhou, China

Renzhong Huang

Hubei Chaozhuo Aviation Technology Co., Ltd., Xiangyang, China

4.1 INTRODUCTION

Cold spray (or cold gas dynamic spray) was initially discovered by Dr. Papyrin's research group during wind tunnel experiments conducted in the mid-1980s in the former Soviet Union [1]. Over the past few decades, as a new alternative to traditional thermal spraying technologies, cold spray is more commonly applied as an advanced coating technology for surface protection or functionalization. In recent years, both academia and industry have paid considerable attention to building 3D bulk components by layer-by-layer addition using the cold spray process. This desire has led to cold spray becoming a new member of the additive manufacturing (AM) family and the term "Cold Spray Additive Manufacturing (CSAM)" has become popular.

In Figure 4.1, in the CSAM process, micron-sized (e.g., 5–50 μm) powder particles are fed into a pressurized (up to 7 MPa) and preheated (up to 1100 °C) propellant gas (usually nitrogen or helium) and accelerated to high velocities (300–1500 m/s) [2]. The highly kinetic in-flight particles are then impacted on the substrate, plastically deform, realign, and finally adhere to form high-quality deposits [2]. Compared to thermal spray and fusion-based

DOI: 10.1201/9781032616025-4

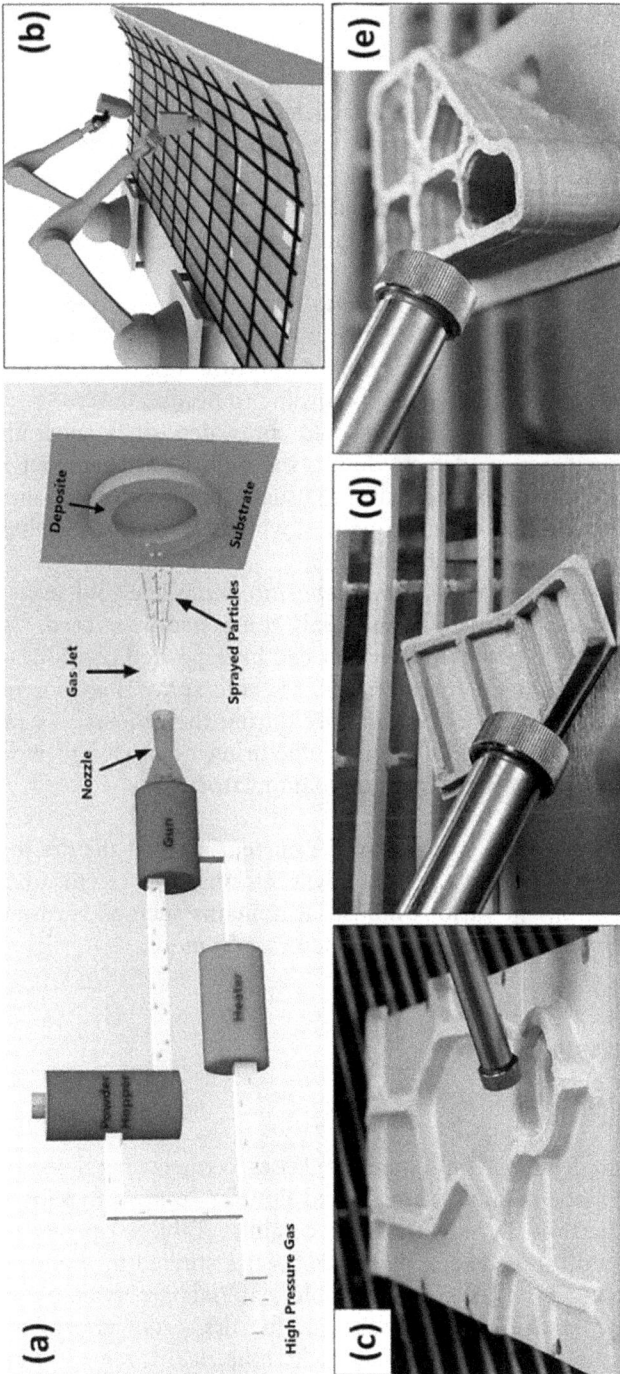

Figure 4.1 Schematics of the CSAM process (a, b) and complex parts/shapes it produced (c, d, e) [5].

AM techniques (e.g., SLM and EBM), the powder particles mainly remain in the solid-state during the CSAM process and therefore exhibit various unique advantages [3, 4]:

(1) *Less heat input*: The powder particles remain in the solid-state due to the low process temperature of CSAM; thus, the feedstock undergoes little phase change and chemical reaction, even in the atmospheric atmosphere. The material deposition process will not generate high thermal stress, thus avoiding significant deformation of the workpiece during preparation.

(2) *A wide range of sprayable materials*: Cold spray can deposit a variety of metals and their alloys or their mixtures, as well as some cermets or metal matrix composites, and the composition of the deposited materials can be flexibly designed according to needs.

(3) *Superior material properties*: Cold spray deposit is generally dense and has almost no defects due to the high particle impact velocity. Therefore, the deposited material usually has good mechanical, thermal, electrical and other properties, which can be comparable to corresponding wrought materials.

(4) *High deposition efficiency and deposition rate*: The cold spray deposition efficiency of metallic materials can generally exceed 90%, and more than 40 kg of metal powder can be deposited per hour.

(5) *Unlimited size of workpiece*: Because cold spray does not require a protective atmosphere or chamber during the process, it can realize surface repair and additive manufacturing of large-size workpieces combined with a high-precision manipulator.

In this chapter, we briefly overview the current state-of-the-art in CSAM, including materials aspects, system advancements, and recent advances of CSAM applications in various industrial domains such as aerospace, biomedical, energy, electronics, semiconductors, and more.

4.2 STATE-OF-THE-ART

4.2.1 Materials

Cold spray can deposit metals, metallic glasses, cermets (metal-ceramic composites), polymers, ceramics, and mixed powders onto different substrate materials such as metals, polymers, and ceramics. Table 4.1 provides a short summary of typical feedstock materials species studied in literature. Note that more detailed lists of cold sprayable materials are available in [6, 7]. In cold spray, hardness is considered as the first-order effect influencing the cold sprayability of a powder particle. Empirically, for <300 HV powders, cold spray can generate deposits of high quality without any thickness

Table 4.1 A short summary of typical feedstock materials in cold spray

Category	Powder material
Pure metal	Al, Cu, Ni, Ag, Ti, Sn, Fe, Cr, Nb, Ta
Alloy	Cu: CuZn, CuCrZr, CuAlFe, CuSn, CuSnP
	Al: AlSi7Mg, Al6061, Al7075, Al2024, AlCu, NiAl, NiAlW
	Ni: In718, In738, In625
	Ti: Ti6Al4V
	Fe: 316L, 304L, FeNi$_{50}$
	Ag: AgPdIn, AgCuTi
	Co: CoCrMo, CoNiCrAlY, CoCrW
	Sn: SnSbCu, SnBi
Mixture	Cu: Cu+W, Cu+SiC, Cu+316L, Cu+Ti, Cu+WC/H13
	Al: Al+Al$_2$O$_3$, Al+SiC, Al+Ti, Al+BN, Al+Fe+Mn
	Fe: Fe+316L, Fe+Al$_2$O$_3$, Fe+Al
	Ag: Ag+W, Ag+graphene, Ag+SiO$_2$
Polymer	UHMWPE, PEEK, HDPE
Ceramic	TiO$_2$, TiN, WC, HA, WC-Co

restrictions; for 300–400 HV powders, cold spray can produce acceptable performance of the deposit, but post-processing may be required, such as HIP or heat treatment; for those powder materials with hardness values higher than 400 HV, cold spray is often difficult and might be used to produce materials with porous structures. The compositional yield of the deposit (difference in composition between the feedstock and deposit) varies with different materials systems [8]. Cold spray has been reported to deposit polymers and ceramics, but the density and adhesion strength are relatively low, which can make thick deposit build-up be difficult. Cold spray can also generate composite materials, by using either composite powders or mixed powders, and the latter is often reported to result in better cold sprayability [9]. For the selection of the substrate, substrate materials having similar deformability or hardness as the powder will often yield higher coating adhesion due to collaborative deformation effect [10].

Al-, Cu-, Ti-, and Ni- alloys are typical important engineering metals in industry and their cold spray performances are thus introduced in detail. Al and its alloys generally have the merits of high ductility, low density, excellent thermal conductivity and electrical conductivity, and good corrosion performance. Cold spray Al and its alloys for repair and restoration of damaged components have attracted significant research efforts from both industry and academia. Al and its alloys generally have low hardness/strength and melting point, and are theoretically easy to cold spray. However, the low deposition efficiency, insufficient cohesion strength, risks of feedstock oxidation and nozzle clogging are potential issues during the cold

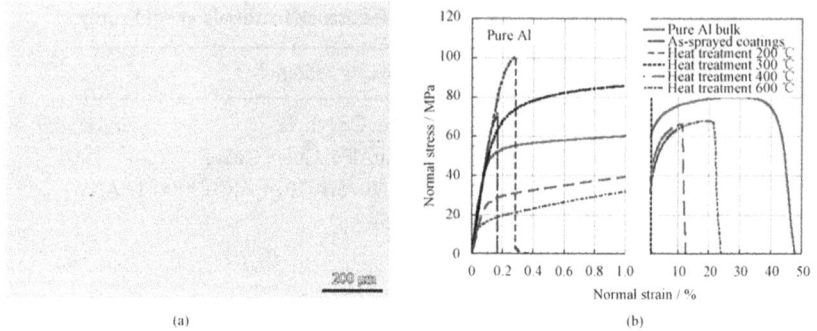

Figure 4.2 Microstructure and stress-strain curves of cold sprayed pure Al: (a) Microstructure of cold sprayed Al coatings and (b) Tensile strengths of cold sprayed Al coatings [11].

spraying of Al and its alloys. Figure 4.2 shows the microstructure and mechanical properties of cold-sprayed Al deposited at optimized process parameters [11]. The cold-sprayed Al is relatively dense, with a porosity level lower than 0.5%. The as-deposited Al has a tensile strength of ~70 MPa, which is higher than the bulk. But the ductility of the Al deposit is poor, with an elongation of only ~0.2%. The mechanical properties of Al deposits can be further enhanced by post-processing such as heat treatment or hot isostatic pressing (HIP). After 200°C heat treatment, tensile strength increases to 100 MPa but the elongation slightly increases to 0.3%. This is due to atomic diffusion between particles to improve the cohesion strength of the deposit. At higher heat treatment temperature of 600°C, the tensile strength decreases but the elongation increases significantly, to 20% [9]. This is because the increase of heat treatment temperature contributes to higher degree of grain recrystallization and growth.

Compared with pure Al, cold-sprayed Al alloys is more commonly studied for industrial purposes. Al-Sn alloys have low modulus and good anti-stick properties and are hence usually used in the automotive industry as sliding bearing materials. Cold-sprayed Al-Sn deposits have low porosity and good mechanical properties [12]. The Al-Si alloy has high mechanical strength, good anti-friction properties and a low thermal expansion coefficient. During cold spraying, the thermal effect of preheated gas leads to the Al-Si alloy deposit having not only an α-Al phase but also fine Si particle precipitates, which are reinforcement phases in the matrix. This means that a higher strength of deposit can be obtained [13]. High-strength Al-Cu alloys are mostly used in the automotive and aerospace industries. The additions of other alloying elements (Ni, Fe, Mg, Ag, etc.) to Al-Cu alloys can generate different intermetallic phases (Al_2Cu, Al_9FeNi, Al_2CuMg, etc.), which further increases their strength. For example, cold-sprayed AA2618 alloy deposits (Al-Cu-Mg-Fe-Ni) are dense and have intermetallic Al_9FeNi precipitations inside the grain and Al_2CuMg in the grain boundary [14].

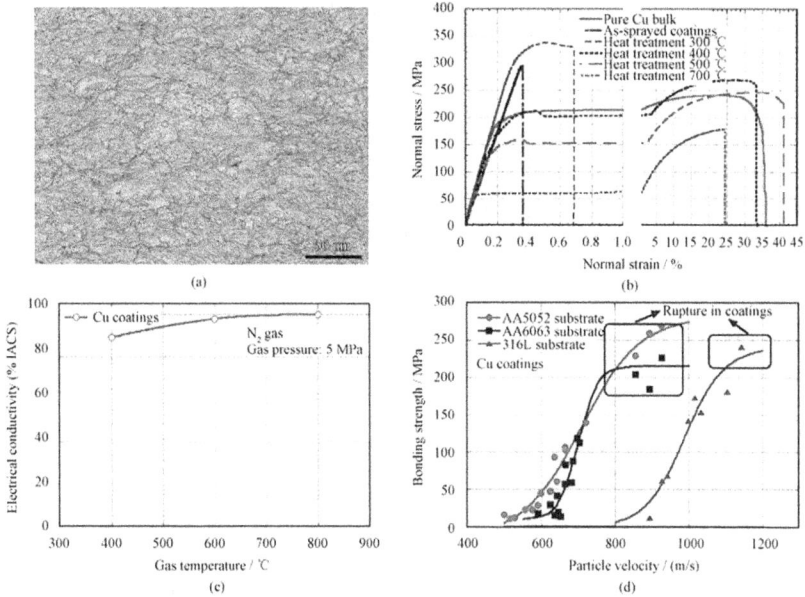

Figure 4.3 (a) Microstructure of cold-sprayed Cu coating, (b) tensile strength of cold-sprayed Cu coatings, (c) electrical conductivity of cold-sprayed Cu coatings, and (d) bonding strength of cold-sprayed Cu coatings [11].

Cu and its alloys have excellent properties of ductility, thermal and electrical conductivity, wear resistance, and corrosion resistance. They are widely used in machinery, electric power, energy, electronics industries. Cu is perhaps the easiest cold-sprayable material and is often used a "model" material to validate fundamental theories. Figure 4.3(a) displays the microstructure of cold-sprayed copper. The copper deposit is almost fully dense with few pores. Figure 4.3(b) displays the stress–strain correlations of cold-sprayed Cu deposits in as-sprayed and heat-treated conditions. Likewise, the as-deposited Cu has higher tensile strength than the bulk (~300 MPa), but its ductility is still low (an elongation of ~0.4%). After heat treatment, the tensile strength of copper deposit decreases and the ductility increases. Note that the as-sprayed Cu deposits have excellent electrical properties with nearly 100% conductivity of the bulk (Figure 4.3(c)). Figure 4.3(d) shows the "adhesion strength–particle velocity" correlations of cold-sprayed Cu deposited on various substrates. The adhesion strengths of Cu on 316L, AA6063, and AA5052 substrates all surpass 200 MPa. As the particle velocity increases, a higher adhesion of Cu deposits can be generated on different substrate materials [15]. It is also noted that the higher the hardness of the substrate, the higher the particle velocity required to establish effective adhesion. Moreover, because of intensive particle deformation and dynamic recrystallization during spray, there are numerous nano grains [16] and nano twins [17] in the copper deposits.

Ti and its alloys have high strength, low density, excellent biocompatibility, and good corrosion resistance. They are commonly used in industries such as chemical, petroleum, aviation, aerospace, automotive, construction, medical, and sports equipment. The cold spraying of Ti generally faces an interesting contradiction of easy deposition and strong adhesion but low density. Ti is also an active material that can easily cause nozzle blockage, and high-strength Ti alloys such as Ti6Al4V might cause wear damage to the cold spray nozzle. Moreover, the anti-friction properties of cold-sprayed Ti and its alloys are poor. Researchers showed that the wear resistance of Ti deposits can be promoted by conducting proper post-treatment or preparing Ti-based composite materials. To deposit dense Ti and its alloys, extreme cold spray parameters must be used (e.g., gas pressure 4–5 MPa, gas temperature 800°C–1100°C, or use expensive helium gas). Studies also show that cold spray of carefully selected mixtures (Ti + Ti6Al4V) can result in dense Ti-based deposits (~1.5% porosity), of which the mechanism is associated with intricate interactions between hard/soft particle interfaces during deposition [18]. Controlling the porosity of Ti and Ti alloys can realize different application purposes, such as porous Ti for biomedical applications and dense Ti for the repair and remanufacture of aerospace components. Figure 4.3 shows the microstructures of cold-sprayed Ti and Ti6Al4V with optimized process parameters. The Ti alloy deposits are dense with no apparent cracks. Further, it was found that the adhesion strength of cold-sprayed Ti6Al4V onto Ti6Al4V substrate can surpass 90 MPa [19] and cohesion strength between Ti6Al4V particles can surpass 350 MPa [20].

In718 is a well-known Ni-based superalloy comprising the main alloying elements of Cr, Ni, Nb, Mo, Al, Ti. In718 exhibits high-strength and excellent high-temperature oxidation and gas corrosion resistance by the precipitations of strengthening phases γ' and γ'' particles as and fine/stable carbides. In718 is broadly applied in aerospace components that must withstand high temperature when in service. Compared with pure Ni, In718 has weaker ductility, and a higher yield strength and strain hardening rate, and is therefore difficult to deposit by cold spray. Ma et al. [21] reported very low porosity (<0.3%) of cold-sprayed In718 through the optimization of nozzle design and parameters (Figure 4.5(a)). Figure 4.5(b) showed that cold-sprayed In718 has a tensile strength >1200 MPa and elongation >9% after heat treatment, which are the highest readings to be found in the public literature. Figure 4.4(c) showed that cold-sprayed In718 exhibited a microhardness of ~590 $HV_{0.3}$ and ~400 $HV_{0.3}$, respectively, before and after heat treatment. Figure 4.5(d) showed that cold-sprayed In718 has an adhesion strength of 400 and 900 MPa, respectively, before and after heat treatment [21]. The tensile strength and adhesion strength of the heat-treated In718 deposit are close to those of as-cast

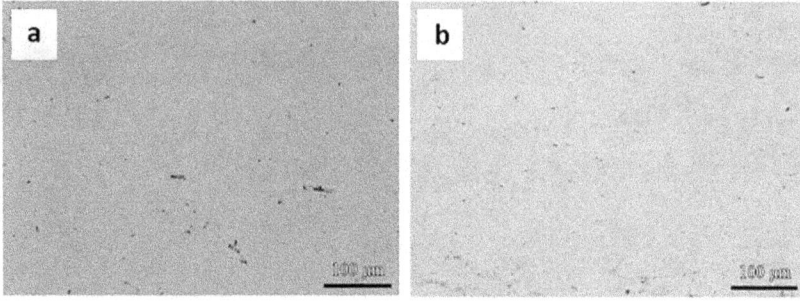

Figure 4.4 Microstructures of cold spray deposited: (a) Ti and (b) Ti6Al4V.

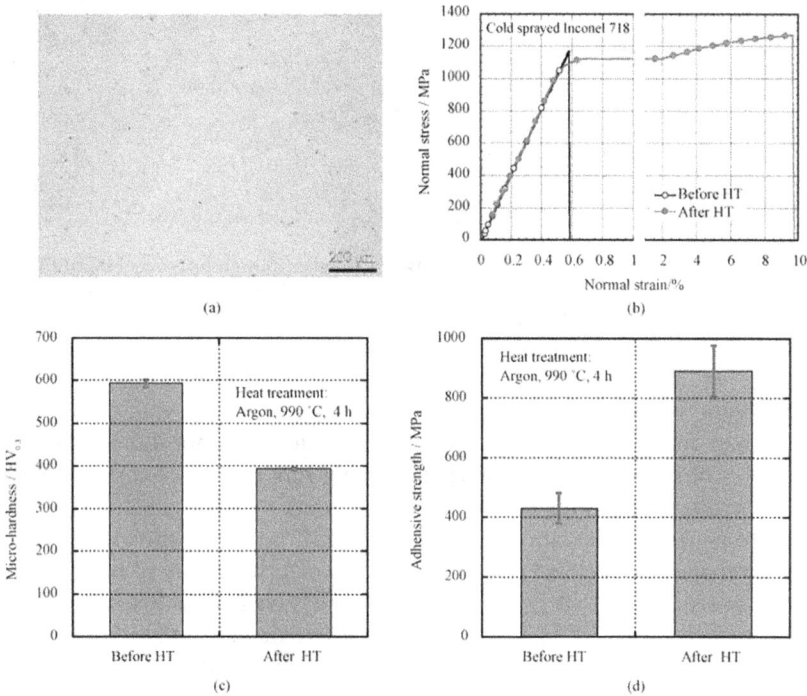

Figure 4.5 (a) Microstructure of cold-sprayed Inconel 718 coating, (b) tensile strength of cold-sprayed Inconel 718 coatings, (c) microhardness of cold-sprayed Inconel 718 coatings, and (d) adhesion strengths of cold-sprayed Inconel 718 coatings [21].

In718. Moreover, cold-sprayed In718 showed higher wear resistance when at high temperatures compared with that at normal temperatures [22]. Based on the excellent overall properties of cold-sprayed In718, it can be potentially used for the surface strengthening of high-temperature parts or the repair of Ni-based superalloy parts.

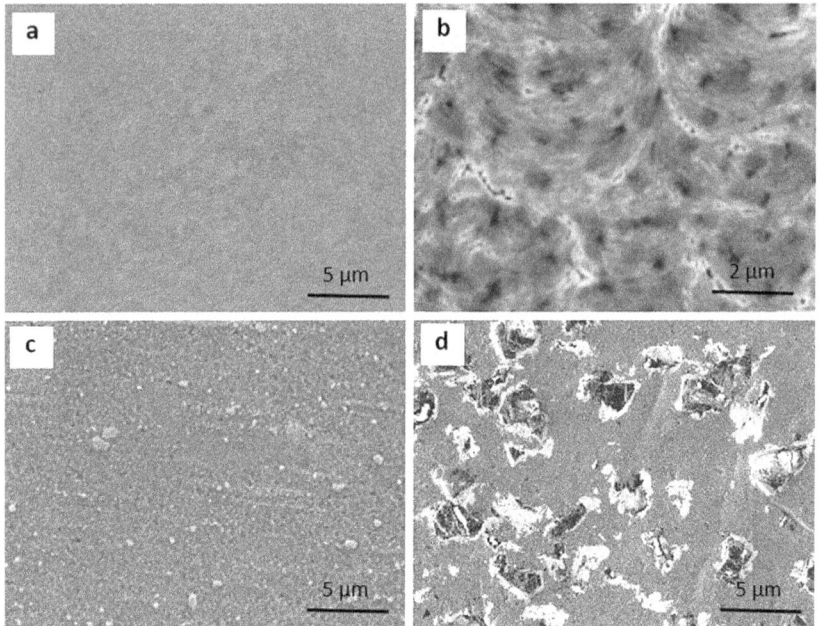

Figure 4.6 Some novel cold spray coating materials: (a) high entropy FeCoNiCrMn alloy [23], (b) Invar alloy [26], (c) Al7075/nano TiB$_2$ composite [27], (d) Al (shell)/diamond (core) [28].

In recent years, new coating materials have emerged which extend the cold spray toward different potential applications. Figure 4.6 shows some examples of the new coating materials being reported in cold spray. Yin et al. [23] cold-sprayed high entropy alloy FeCoNiCrMn using helium gas cold spray to generate thick and dense coatings with no segregation of elements. High-entropy alloys (HEAs), a recent member of the family of metal alloys, were discovered in 2004 by Yeh et al. [24] and have been substantially studied in recent years. HEAs are comprised of five or more main alloying elements in the same or similar mole fractions. Owing to the unique compositions, HEAs exhibit microstructures and properties of their main alloying elements as a whole rather than of the individual constituent element [24]. HEAs exhibit excellent properties, such as mechanical properties, corrosion resistance, wear resistance, and oxidation resistance, when compared with traditional metal alloys [25], and can be potentially applied in many industrial domains such as nuclear power, aerospace, and shipbuilding. Chen et al. [26] cold-sprayed Invar 36 alloy deposits with a high-density, low thermal-expansion coefficient, as well as good mechanical and thermal performance. Xie et al. [27] cold spray deposited Al7075 powder reinforced with nano TiB$_2$. After heat treatment, the Al7075 composite deposit can achieve simultaneous enhancements in ductility and strength compared with Al7075 bulk. Yin et al. [28] reported Al matrix composites reinforced with diamond (DMMCs) developed by the cold spraying of core-shell structure powders. The use of core-shell structure powder (core-diamond, shell-Al) can retain a higher fraction of intact diamond, leading to better wear resistance when compared with mixed powders.

4.2.2 Applications

4.2.2.1 Remanufacture & repair

Traditional techniques such as thermal spray, welding, and laser deposition are currently the most used methods to repair damaged parts which have a high value. However, these conventional techniques could cause unwanted severe defects such as thermal distortion and delamination, and oxidation inclusions due to the redundant heat input during process, which therefore greatly restrict their repair capabilities. Moreover, it is difficult for parts with thin wall structures and complex shapes to be repaired using the above techniques, which are therefore scrapped resulting in a tremendous waste of materials. The distinct low-temperature, solid-state process characteristics of the cold spray technique allow it to avoid high temperature-induced defects. It has shown great potentials in terms of the repair and remanufacture of naval, aircraft, aerospace, and automotive parts. Cold spray additive manufacturing technology has been favored and extensively studied by U.S. Department of Defense and U.S. Army Research Laboratory (ARL). The U.S. Department of Defense, for example, has provided long-term substantial research funding for cold spray repair of worn parts.

Figure 4.7 shows the fretting damage on the surface of A357 as-cast Al alloy parts of F/A-18E/F fighter. The cold spray technique was used to dimensionally and functionally restore the part [29]. Figure 4.8 shows another example in which cold spray was used to repair Seahawk helicopter parts, which could save around 35–50% when compared with the cost of replacement [30]. In addition, the U.S. Army Research Laboratory worked with General Electric and Moog to repair the GE T700 Front Frame Housing through using cold spray technology. The front frame is made of cast C355

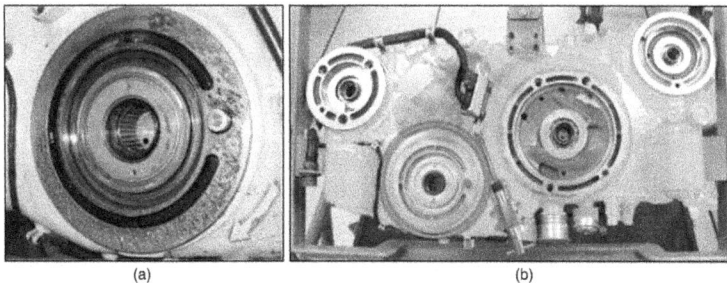

(a)   (b)

Figure 4.7 Cold spray repair of as-cast Al parts in F/A-18E/F fighter aircrafts: (a) Fretting damage and (b) after repair [32].

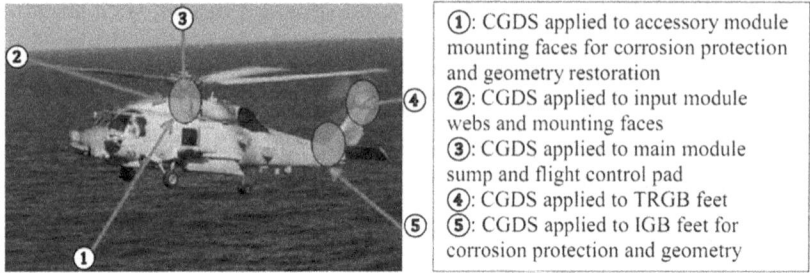

①: CGDS applied to accessory module mounting faces for corrosion protection and geometry restoration
②: CGDS applied to input module webs and mounting faces
③: CGDS applied to main module sump and flight control pad
④: CGDS applied to TRGB feet
⑤: CGDS applied to IGB feet for corrosion protection and geometry

Figure 4.8 Cold spray repair of Seahawk helicopter parts [30].

As-received

Cold sprayed

Final machined

Anodized

Figure 4.9 Cold spray repair of T700 Front Frame [33].

Al alloy and 6061 Al alloy was selected as the repair material. Figure 4.9 shows the T700 Front Frame from as-received to the as-finished product after procedures of cold spray, machining, and anodizing [31].

In addition, T. Stamey [34] reported the cold-spray repair of marine pump casing. The base material of the pump is a kind of widely used tin-bronze alloy, which is used in wear and corrosion conditions. In general, the tin-bronze alloy is not easy to repair through welding, meaning that cold spraying was chosen as the best option for repairing. A corroded part of a naval warship was also repaired by Moog using high-pressure cold spray and was returned for future service [31]. The part was made of Al 6061 alloy and contained significant corrosion damage that otherwise would have resulted in it being scrapped. The part after cold spray repair and subsequent machining/

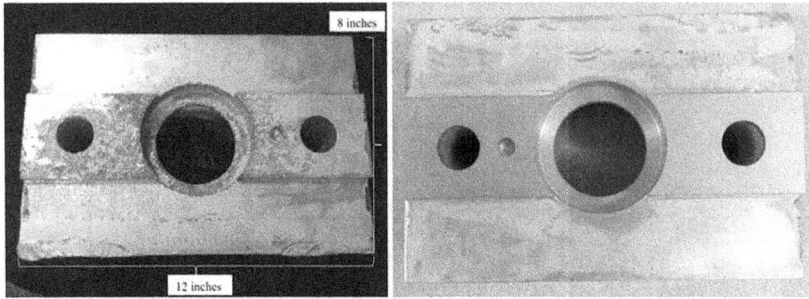

Figure 4.10 Corroded vessel actuator before and after repair by cold spray [33].

anodizing is shown in Figure 4.10. The repaired actuator was back in service, alleviating the lengthy lead times and the large expense involved in a replacement purchase.

In addition, many other research entities or manufacturing companies, such as Boeing, Pratt & Whitney, Honeywell, GE, Safran, Rolls-Royce, Airbus, TWI among others, have also made great efforts in pushing forward cold spray to repair high-value worn parts. However, the current implementations of cold spray technology as a repair technique are mainly on non-structural components. The structural repair by cold spray still present a large number of challenges. Significant improvements in adhesion/cohesion strength and ductility of cold-sprayed deposits remain as future challenges. Thus, suitable in-situ or post-process treatments on the cold-sprayed deposits should be further adopted.

4.2.2.2 Biomedical

Cold spray technology is also a method to fabricate coronary stents with superior mechanical and degradation properties. In one study, a McGill University research group, through the cold spraying of mechanically mixed 316L/Fe powders, subtractive machining and vacuum heat treatment, produced novel amalgamate stents [35]. The biodegradable stent is patented and is currently under clinical trials. The concept of 316L/Fe bi-material with different electric potentials introduces the microgalvanic effect and therefore accelerates the overall corrosion rates. By adjusting the relative mixing ratios of 316L and Fe in the feedstock materials, the corrosion rate can be varied and controlled, which is particularly beneficial for biodegradable scaffolds, as obtaining controlled degradation is critical to their successful clinical implementation. The grain size of a produced 316L/Fe stent is about 10 µm, which reveals promising mechanical properties such as fatigue and tensile strength. Figure 4.11 shows the whole stent fabrication process.

Overall, fundamental issues of cold-sprayed biomedical deposits, such as material selection, coating processing, coating formation, microstructure–property relationships, and biological properties, are widely understood.

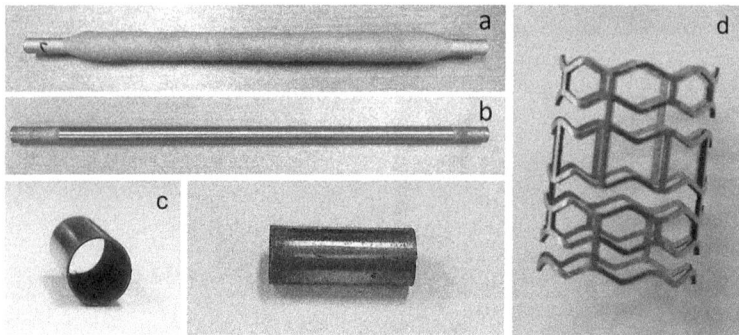

Figure 4.11 Stent fabrication process: (a) cold spray on a cylindrical substrate, (b) grind and polish the surface, (d) electric discharge machining to separate deposits, (e) finished stent after laser cutting and electropolishing [36].

However, most cold spray biomedical deposits are still at the preclinical testing stage. Not enough is known about the in-vitro and in-vivo behavior of cold-sprayed biomedical deposits at different levels. The fabrication of advanced biomedical deposits and biomedical devices using cold spray additive manufacturing technology requires further exploration.

4.2.2.3 Semiconductor & Electronics

Cold spray additive manufacturing technology can also be used in electronics and semiconductor production. There is an extremely wide range of materials, including Ag, Ta, Cu, Ti, Nb and other pure metal materials, that can be used in the sputtering process for electronic coatings. These coatings include solar cell coatings, semiconductors, glass coatings, etc. Most modern electronics contain basic components produced from Ta sputtering targets. These include memory chips, microchips, flat panel displays, printheads and more. Sputtering targets are used to produce Low-E glass, which is often used in building construction to control light, save energy costs and for aesthetic reasons. The demand for renewable energy is significantly increasing due to global climate change. The third-generation thin-film solar cells are also fabricated by sputtering coating technology.

Figure 4.12 shows a large-scale rotary sputtering target through cold spray additive manufacturing and post-processing. Cold spray additive manufacturing can produce thick metal deposits in a very effective and efficient manner. The adhesion strength of cold-sprayed deposits and production efficiency are much higher than those of traditional manufacturing techniques. The oxygen content and porosity are also much lower. In addition, its grain size, on average, is much smaller than that of the as-cast material. Thus, large-scale metallic sputtering targets directly formed by cold spray additive manufacturing have been used in the electronic information

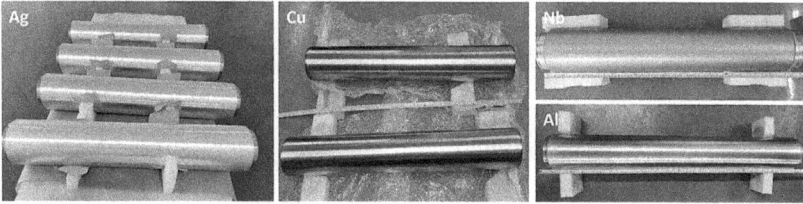

Figure 4.12 CSAM for sputtering target applications.

industry, including in the manufacture of liquid crystal displays, information storage, laser memory, integrated circuits, glass coatings, electronic control devices and other fields. Similarly, cold spray can also produce large-scale rotary sputtering targets made of other metals, such as Ni, Ti, Ta, Mo, etc. However, the current challenge is how to efficiently manufacture low-oxidation, high-purity sputtering targets by the cold spray additive manufacturing method. The development of low-oxygen metal powders and proper powder handling are both directions worth exploring.

4.2.2.4 Energy & power

Not only can cold spray additive manufacturing avoid environmental pollution problems compared with traditional electroplating, but, more importantly, it has the advantages of excellent coating performance and no limitation of coating thickness, making it especially suitable for depositing Ni, Ag, Cu coatings on electrodes, printing rollers, batteries, etc. Figure 4.13 shows a gravure-printing Cu roller. The Cu coating is deposited by cold spray additive manufacturing. Compared to the traditional electroplating process, cold spray additive manufacturing is both more efficient and less polluting, and can meet the requirements of thick plating to ensure that printed products have better 3D effects.

As shown in Figure 4.14, another application example is the cold-sprayed copper deposits on used nuclear fuel containers being developed in Canada for the storage of used nuclear fuel. In this instance the vessel consists of a steel vessel with a copper coating on the surface for corrosion protection. The copper deposit is produced by the cold spray process. Since the copper

Figure 4.13 CSAM of thick Cu deposits on printing roll.

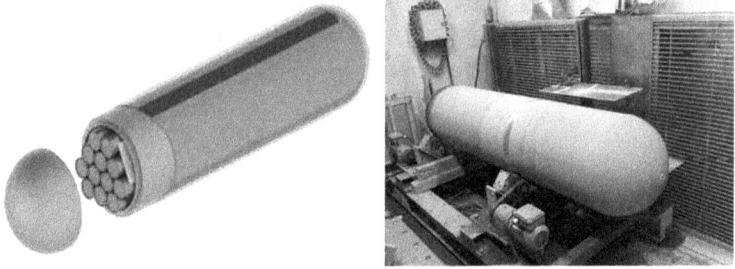

Figure 4.14 CSAM of Cu deposits on steel vessels for nuclear fuel storage [30].

deposit is not a structural component, the strength of the copper is less critical. Local heat treatment is conducted to restore the ductility of cold-sprayed copper. Copper deposition by cold spray could probably provide a great solution for long-term storage of used nuclear fuel waste.

4.2.2.5 Aerospace & Astronautics

Ti, Ni, and Cu alloys are widely used in astronautics and aerospace industries owing to the extreme demands of high strength at elevated temperatures and the high heat dissipation capabilities. Therefore, cold spray is a natural fit AM alternative to fabricate components of these metals, and, in particular, is suitable for situations where a dissimilar materials combination is required. Figure 4.15 shows typical examples of aerospace components

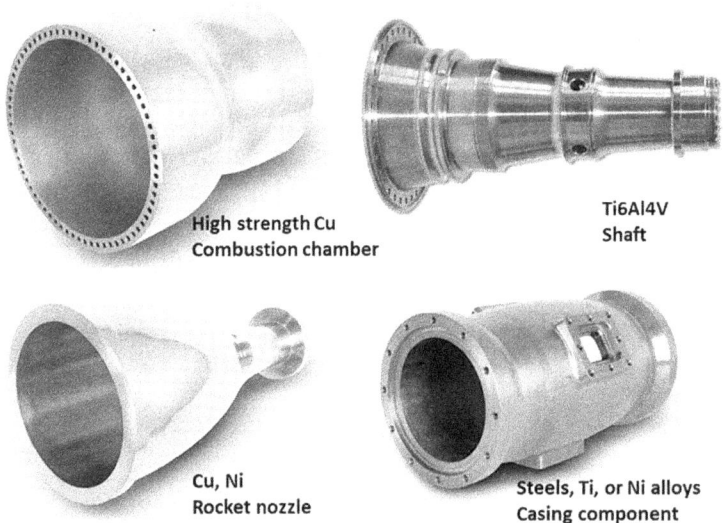

High strength Cu
Combustion chamber

Ti6Al4V
Shaft

Cu, Ni
Rocket nozzle

Steels, Ti, or Ni alloys
Casing component

Figure 4.15 CSAM of typical aerospace components [37].

made by Impact Innovation Inc. through cold spray additive manufacturing [37]. In another project a NASA project team recently conducted hot-fire testing of CSAM lightweight combustors and nozzles [38]. Compared with the traditional subtractive methods, which require complex and lengthy machining operations, cold spray additive manufacturing produces a near-net-shape shape and the components only require minimized post-processing. Therefore, the production time is greatly reduced, and the production efficiency is improved, meaning that significant material savings can be realized.

4.3 CSAM PROCESS CHARACTERISTICS

4.3.1 CSAM system

CSAM is a recent 3D printing technology developed based on cold spray. Its basic principle is still to use cold spray technology for free manufacturing layer by layer. As with cold spraying, the CSAM system is mainly composed of gas heater, high pressure gas, spray gun, control system, powder feeder, robot, and peripheral equipment [39]. The high pressure gas can be air, nitrogen or helium [7, 40]. The substrate can be stationary, and the robot drives the spray gun to move to deposit a plane structure or a curved surface structure; it can also be moving, usually rotating, and the spray gun remains stationary to obtain a rotating shaft structure. The two can also be combined to create complex structures [41–43]. The comparisons of CSAM to other manufacturing technologies are summarized in Table 4.2.

4.3.2 CSAM process

Like other AM techniques, the entire CSAM process can be divided into three steps: pre-spray, spray and post-spray. The three steps can be described by Figure 4.16 and are thus discussed individually below.

Table 4.2 Comparisons of CSAM to other manufacturing technologies

Metric	CSAM	Subtractive manufacturing	3D printing-based laser
Resolution	Low	High	0.1~100 mm
Size	Large	Unlimited	Small to medium
Speed	Fast	Middle	Slow
Complexity	Low	Low	High
Whether post-processing is required	Yes	No	Yes
Anisotropy of mechanical properties	Must be considered	Consistent	Need to consider

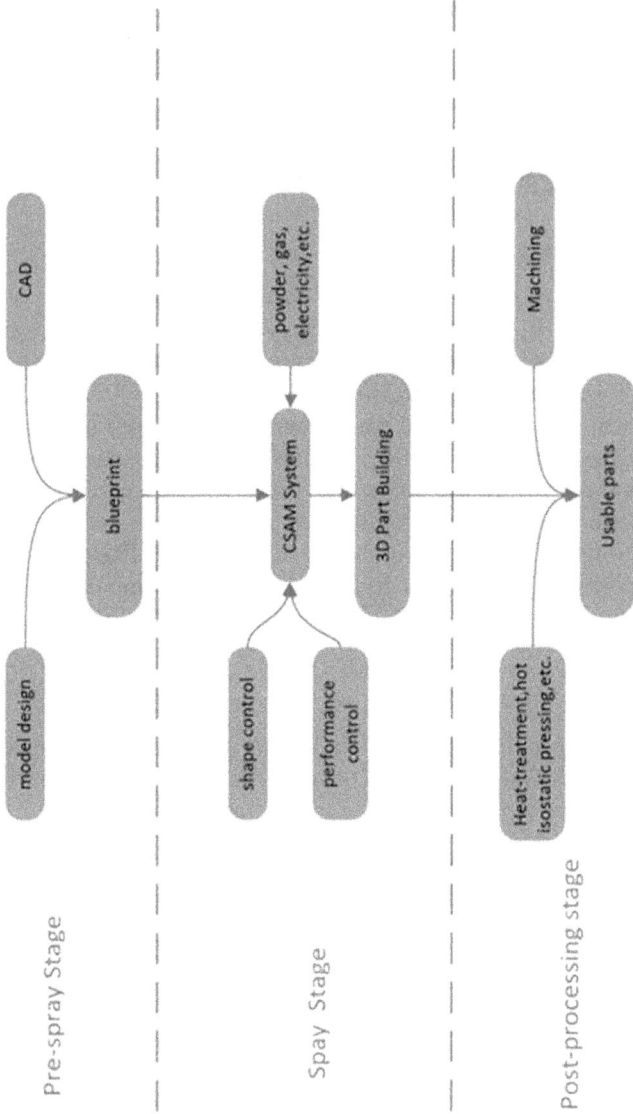

Figure 4.16 Essential steps in CSAM.

4.3.2.1 Pre-spray

3D models used by CSAM are usually designed and modeled based on actual samples. After the digital CAD model is obtained, it is sliced, and the nozzle motion path is planned and created. Since it is difficult to use the support structure for CSAM, the single-layer spraying through the nozzle usually needs to be designed into a curved structure to avoid it.

However, although it is possible to compile directly from the G code currently used in mainstream 3D printing into a cold spray robot, there is no effective way to plan and create a path for 3D manufacturing owing to the characteristics of CSAM. However, there will still be many incompatibilities in the executed path program. Therefore, it is urgent to develop CSAM path planning algorithms and dedicated software.

4.3.2.2 Spray process

During the spray process, the CSAM system will control gas parameters (type, pressure, and temperature), nozzle parameters (traverse speed, standoff distance, spray angle/trajectory, scanning step) and powder feeder parameters (powder feeding rate) to control the quality of deposit.

Increasing the in-flight velocity of feedstock powders can increase deposition efficiency, increase bonding strength, and lower porosity. Generally, higher gas pressures and temperatures increase the in-flight velocity of powder particles; however, it may clog the nozzle [44]. The most effective way to increase the in-flight velocity of powder particles is to use helium instead of air and nitrogen as the propellant gas, but the cost will increase greatly [45].

One of the advantages of CSAM is that multiple layers of different materials or gradients can be fabricated [7, 46] to obtain an excellent bulk-like structure. The replacement of nozzles and powders also needs to be considered in the spraying process for this purpose.

The residual stress of cold spray is also an important factor for the change of shape. Usually, the residual stress of cold spraying is mainly compressive stress. If a flat plate is sprayed, the stress gradient will cause inward bending. In this regard, different powder residual stresses exhibit inconsistent performance on the deformation of the workpiece, and copper is more serious than other metals such as aluminum, iron, etc. Studies have shown that increasing the coating thickness will increase the residual stress, which is more likely to cause deformation or distortion of the workpiece [47, 48]. The geometry control is the most important point in the CSAM process, which will be highlighted in the next section.

In addition to shape control, mechanical properties of the deposits need to be further considered. Compared with other AM technologies and subtractive manufacturing technologies, the tensile strength of the deposit is lower, which therefore needs to be considered for the sprayed workpiece and components.

4.3.2.3 Post-spray

The largest difference between CSAM and traditional additive manufacturing methods is that the properties of CSAM deposits have poor plasticity—meaning that they exhibit almost no ductility [49]. Therefore, post-treatment is essential to further improve the mechanical properties of the CSAM deposits. With the exception of mechanical properties, other properties, such as electrical conductivity [11] and thermal conductivity [50], are generally lower than machined parts.

Heat treatment is a common and versatile method to improve properties of the deposits. Heat treatment can release residual stress, reduce internal defects, and significantly improve tensile strength and ductility [49, 51–53]. Longer annealing times and higher annealing temperatures could result in more significant property improvements. Under an appropriate annealing scheme, the tensile strength and elongation of annealed Cu deposits can achieve the same level as bulk Cu [54]. Heat treatment will also cause certain deformation and distortion of the workpiece, which should be considered.

The work-hardened deposits are softened due to recrystallization that occurs during heat treatment. Therefore, the annealed deposits are usually softer than as-sprayed deposits [55, 56]. In addition, by "healing" inter-particle boundaries, barriers to current and heat transport are removed. Therefore, annealed CSAM deposits could exhibit better thermal [50] and electrical conductivity [57].

CSAM typically produces very porous and rough surfaces that are not suitable for direct use. Machining is an essential step of CSAM. The workpiece is usually treated by direct machining, including alternate spraying based on cold spray and machining, so as to obtain a better part surface [58]. For the preparation methods of complex structures, it is a better method to adopt a variety of additive and subtractive manufacturing methods. Machining is an essential step of CSAM.

Hot-isostatic-pressing (HIP) is also a good technique to improve the properties of the deposit, which helps to reduce porosity and increase tensile strength. Petrovskiy [59] used HIP to study microstructures and tensile strengths of cold spray pure Ti deposits. Studies have shown that the pores are reduced, and the tensile strength is improved.

4.3.3 CSAM process control

Compared to selective laser melting (SLM) or other fusion-based additive manufacturing techniques, cold spray is characterized by high-deposition efficiency, low residual stress, low oxidation, and preserves the original microstructure and mechanical properties of feedstock materials. After the spray gun is installed on the robot, the nozzle can obtain more degrees of freedom, which provides the possibility for the preparation of complex shapes and free-forming. Like other AM techniques, the key point of CSAM is shape control.

4.3.3.1 Nozzle

The spatial distribution of the powder particles leaving the nozzle at high speed is very important to the shape and size of the deposited coating. Usually, the nozzle outlet diameter used by CSAM is 4~10 mm, and the smallest width of spray trajectory is determined by the outlet diameter of the nozzle [58]. Therefore, the width of the coating track on the substrate is > 4 mm, so it is difficult for CSAM to manufacture small-sized parts or parts with tiny detailed features. As shown in Figure 4.17, the velocity and spatial distribution of powder particles in the nozzle are not uniform. In the case of axisymmetric circular nozzles, the distribution of particles at the outlet is almost close to Gaussian distribution [60, 61].

During the spraying process, the particle velocity and distribution in the middle of the particle beam are higher than those in other areas. With the same trajectory spraying multiple times, powder particles will accumulate

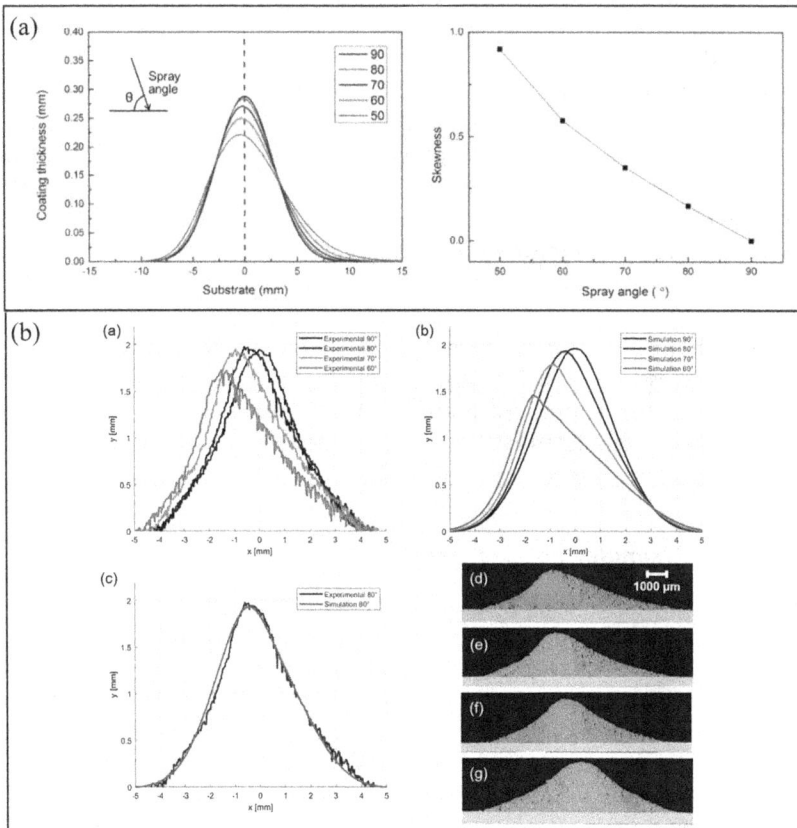

Figure 4.17 (a) Coating profiles and skewnesses at different spray angles [62], (b) the profiles and obtained deposits for spray angle varying from 90° to 60° [63].

rapidly in the center area of the coating. For angle-sensitive processes, an impact angle of 90° can reach the highest deposition efficiency [64, 65]. For particles with small spray angles, their velocity component perpendicular to the substrate cannot reach the critical velocity and thus particles bounce off without forming deposits [66]. When using axisymmetric nozzles, the profile shape can be often described by a Gaussian function for a small number of layers and relatively high lateral velocities of the nozzle, and it gradually evolves into a triangular profile if lower velocities or numbers are used [67, 68]. At the same time, the slowly rising deposit morphology in the middle is also not beneficial to the deposition of subsequent particles. Thus, spray coating will eventually form a coating on the substrate with a cross-section approaching a triangular profile trajectory.

To obtain a more uniform single-track deposit, elliptical and rectangular nozzles were studied. Tabbara et al. [61] studied the effect of nozzle shape on the exit velocity and spatial distribution of the particles. The results have shown that powder particles acquire radial velocity components in the expansion section of Laval nozzle, and it is difficult to produce a uniformly distributed spray trajectory.

The degree of powder dispersion and the Gaussian distribution theorem, the discrete control model, and the cross-sectional geometry of the cold spray deposit (the degree of powder dispersion) are affected by many factors, including nozzle dimension, spray process parameters (temperature and pressure), spraying distance, selection of different substrates and powder materials. Therefore, it is usually necessary to establish a different spraying database for each powder and process parameter to meet different needs.

However, no matter how the nozzle shape is designed, in the long-term spraying process, the powder impact angle and deposition efficiency in different areas of the coating track are constantly changing, so that there will inevitably be accumulated errors in the long-term spraying process, which will further reduce the accuracy and affect the control of CSAM shape.

4.3.3.2 Track simulation spray strategy

Whether CSAM can become an additive manufacturing method for fabricating complex shapes depends on the planning of the deposition trajectory. Control of deposit tracks typically includes spatial resolution control and minimal fabrication units (e.g., individual spray track shape). The minimum manufacturing unit mainly depends on the nozzle dimension. The nozzle restricts the minimum width of each track to the nozzle outlet diameter. To solve the problem of too wide a track and improving the cross-sectional resolution of the coating, Sova et al. [58] designed small nozzles to reduce the spray width and improve the accuracy of CSAM, and a spray track width of <1 mm was obtained. However, it is difficult to obtain dense coatings due to the short acceleration route of powder particles.

To improve the control accuracy of a single track, a series of studies have developed different prediction models for single-track deposition. Djurić et al. [69], through inversely transforming the nonlinearity into a boundary value problem, developed a deposition model which could simulate the spatial mass distribution of powder particles (e.g., deposition efficiency) in the nozzle. Duncan et al. [70] also adopted a numerical model developed by Djurić to optimize path trajectories during spray painting, which utilized the sampling theory to transfer problem from time domain to spatial frequency domain [62].

In CSAM, in addition to temperature, pressure and nozzle dimension, the sprayed coating cross-section is also affected by spraying distance and spraying angle. It is difficult to use path planning algorithms of subtractive manufacturing and other AM technologies to create the spraying trajectory of CSAM. Some typical spraying strategies have been proposed, including multiple materials, embedded 3D architectural measurement, complex spraying trajectories can be constructed using five-axis or six-axis robots, and vertical surfaces having complex morphologies or high spatial resolution can be obtained.

Lynch [41] has achieved remarkable results in the combination of properties of cold spray deposits and topology optimization ideas to design and manufacture a scaffold structure. As shown in Figure 4.18, a reliable guiding process is proposed: from designing 3D structure to using topology optimization techniques and controlling of spray paths, and finally to the further processing of the obtained blank to form the shape of the final stent.

Lynch [41] also revealed the feasibility of 3D rapid prototyping of complex shapes. Through the geometry of some specific shapes and features (smooth transitions between curvatures, planar features, truss structures, etc.), the material of the finished surface is produced by using post-processing to remove excess deposits, although some improvements are required in order to use the cold spray operation process.

4.3.3.3 Closed loop control and online monitoring for CSAM

Usually, path planning and simulation can help researchers design parts before spraying, and theoretically obtain expected part dimension and performance. In the actual spraying process, various errors and disturbances will be encountered. The factors affecting the shape and geometric accuracy include stability of gas pressure and temperature, material, process parameters of the powder and last spray, gun trajectory, impact angle, robot errors, and environmental factors. The variety of different materials and process parameters makes it difficult to establish accurate spray performance for each material and process. Even with a lot of upfront research on a material, some problems can be encountered during the spray process. The best way to deal with these errors is to implement a compensation strategy through closed-loop control.

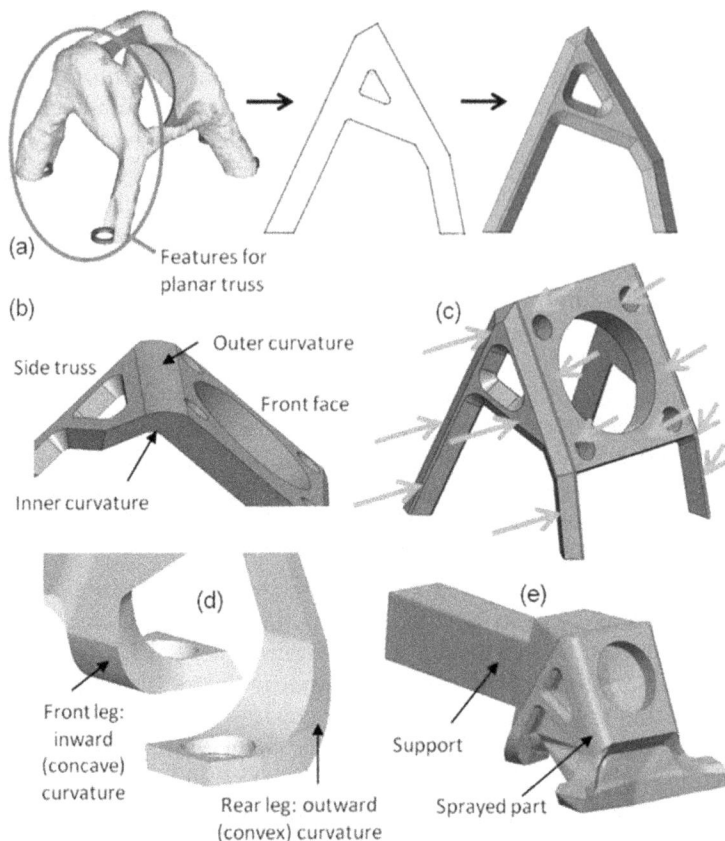

Figure 4.18 CSAM design based on topology [30].

Introducing perception and decision-making learning methods into cold spraying to realize feedback control in the process of cold spraying and realize a high-precision closed-loop control system is the best way to solve the low control accuracy of CSAM.

For CSAM, some feasible online monitoring methods include monitoring the flight speed and temperature of powder particles and measuring the temperature and shape of the coating. For example, a high-speed 3D laser profiler is used to detect the coating profile and compare it with the simulation model to correct the spraying trajectory and process parameters in the real time, as shown in Figure 4.19. The LERMPS laboratory in France have begun to conduct preliminary feasibility studies on the CSAM line monitoring technology [60]. To be able to detect the coating deposition process online, a high-speed 3D laser profiler was introduced into the system to scan the coating profile.

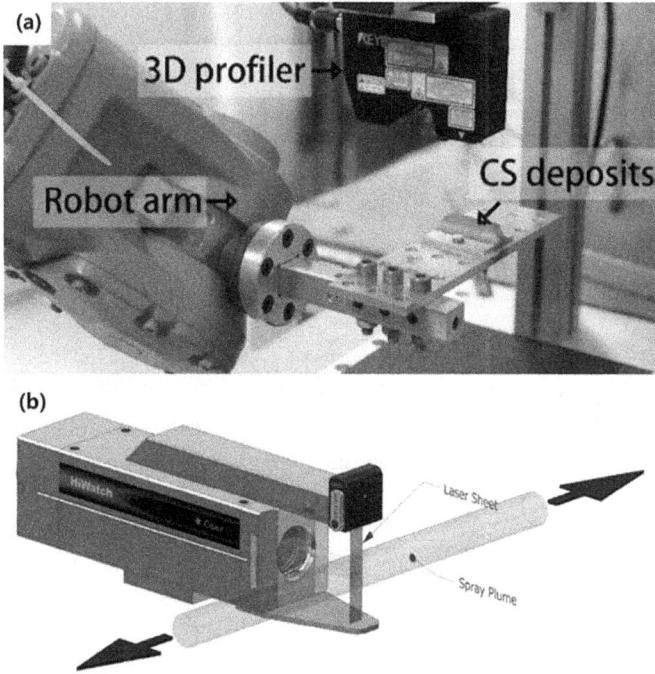

Figure 4.19 (a) Online 3D profiler monitor and (b) particle velocity monitoring [71].

Nault et al. [72] put forward a novel toolpath optimization algorithm for the fabrication of arbitrary convex deposit shapes. The algorithm compensates for local deviations in the deposit thickness due to the part geometry by correcting toolpath velocity. The model is calibrated experimentally, including line spray and particle velocimetry measurements for closed-loop control.

4.3.3.4 Deep learning for CSAM shape control

Artificial neural network (ANN) has the ability of self-learning, self-adaptation, self-organization, the strong approximation of nonlinear function, and strong fault tolerance. It can realize functions such as image recognition, simulation, prediction and control. It is a powerful tool to deal with nonlinear systems.

The method based on deep learning is currently a better solution to the spraying characteristics of various powders and positional process parameters. At present, ANN modeling has demonstrated its powerful functions in other additive manufacturing technologies. In CSAM is still in the early research stage.

The single-trajectory spraying process adopts neural network design with very important characteristics; the application of artificial neural network modeling in prediction is greatly limited to the characteristics of key geometry only for additive manufacturing, such as height and width. Cheng et al. [73] improved the variations in deposit quality during cold spray molecular dynamics simulations through the development of a deep convolutional neural network-assisted optimization method. Ikeuchi et al. [74, 75] used a neural network model to establish the modeling, and the results showed that the neural network modeling method is very suitable for the prediction of the cold spray profile. Compared with the traditional Gaussian modeling method, the model prediction of the contour edge area is more accurate.

The accurate modeling of asymmetric and symmetric monorail profiles with a high degree of geometric detail will help improve the CSAM geometric control, allowing it to produce more consistent and complex components with minimized post machining.

4.4 CONCLUSIONS & FUTURE PERSPECTIVES

This chapter briefly overviewed the state-of-the-art in CSAM. CSAM offers entirely new opportunities for the solid-state fabrication of large-scale 3D metallic materials. Compared with traditional fusion-based additive manufacturing techniques, CSAM can be deposited on a large scale in a short period of time while keeping the feedstock materials properties unchanged. There are many unique advantages of CSAM in terms of individual components. CSAM has been used to deposit various kinds of materials, resulting in thick, dense deposits showing high deposition efficiency. Thus, increasing attention from both academia and industry has been paid to this rising AM technology.

However, there are also some challenges ahead to overcome for further practical applications of cold spray additive manufacturing. First and foremost, the relatively large diameter of the nozzle exit severely limits the resolution of the printed parts, which essentially determines the characteristics of cold spray additive manufacturing. Unlike SLM or EBEM processes, CSAM is not suitable for the fabrication of small/thin walled/complex-shaped/internal structures. Nevertheless, CSAM has its niche in fabricating large-scale and simple-shape components, such as cylindrical components, which presents cost-saving and high-efficiency advantages.

Another challenge of CSAM is the precise control of the shape of the printed structures and the toolpath. Numerical simulation is an effective method to visualize the layer-by-layer deposition process and predict the thickness and morphology of the deposition layers, which could greatly assist in the optimization of cold spray parameters and strategies. Besides, machine learning modules and process monitoring can be combined with CSAM to make this process smarter and more intelligent.

In addition, due to the solid-state bonding mechanisms and the non-fusion principle of CSAM, the deposits usually have lower ductility and strength than corresponding bulk materials, meaning that suitable in-situ or post-processing methods need to be explored. In-situ treatment methods include incorporating laser or shot peening or heating techniques during the CSAM process, which could further enhance the qualities of CSAM parts. Post-process treatments, such as heat treatment and HIP, can be used to improve the mechanical properties of cold spray additive manufacturing parts. By employing appropriate in-situ or post-processing, the future of manufacturing structural parts by CSAM is promising.

Thus, efforts are still required to further develop CSAM, so that an increasing number of types of materials could be used for CSAM and large-scale complex-structure components with excellent properties can be fabricated through the process. If this occurs, the applications of CSAM technology could be expanded to the aerospace, aircraft, oil & gas, defense and more industries.

REFERENCES

1. A. Papyrin, Cold spray technology, *Advanced Materials & Processes*, **159**(9), 49–49 (2001).
2. T. Schmidt, F. Gärtner, H. Assadi, H. Kreye, Development of a generalized parameter window for cold spray deposition, *Acta Materialia*, **54**(3), 729–742 (2006).
3. W. Sun, A. W.-Y. Tan, K. Wu, S. Yin, X. Yang, I. Marinescu, E. Liu, Post-process treatments on supersonic cold sprayed coatings: A review, *Coatings*, **10**(2), 123 (2019).
4. R. Huang, H. Fukanuma, *Future trends in cold spray techniques, Future Development of Thermal Spray Coatingsed*, Elsevier, 2015, pp. 143–162.
5. Y. Zou, Cold spray additive manufacturing: Microstructure evolution and bonding features, *Accounts of Materials Research*, **2**(11), 1071–1081 (2021).
6. H. Assadi, H. Kreye, F. Gärtner, T. Klassen, Cold spraying – A materials perspective, *Acta Materialia*, **116**, 382–407 (2016).
7. A. Moridi, S.M. Hassani-Gangaraj, M. Guagliano, M. Dao, Cold spray coating: Review of material systems and future perspectives, *Surface Engineering*, **30**(6), 369–395 (2014).
8. X. Chu, H. Che, C. Teng, P. Vo, S. Yue, A multiple particle arrangement model to understand cold spray characteristics of bimodal size 316L/Fe powder mixtures, *Surface and Coatings Technology*, **381**, 125137 (2020).
9. X. Chu, H. Che, C. Teng, P. Vo, S. Yue, Understanding particle–particle interactions from deposition efficiencies in cold spray of mixed Fe/316L powders with different particle size combinations, *Journal of Thermal Spray Technology*, **29**(3), 413–422 (2019).
10. K. Wu, W. Sun, A. W.-Y. Tan, S. C. Tan, E. Liu, W. Zhou, High temperature oxidation and oxychlorination behaviors of cold sprayed Inconel 718 deposits at 700 °C, *Corrosion Science*, **207**, 110536 (2022).

11. R. Huang, M. Sone, W. Ma, H. Fukanuma, The effects of heat treatment on the mechanical properties of cold-sprayed coatings, *Surface and Coatings Technology*, **261**, 278–288 (2015).

12. X.-J. Ning, J.-H. Jang, H.-J. Kim, C.-J. Li, C. Lee, Cold spraying of Al–Sn binary alloy: Coating characteristics and particle bonding features, *Surface and Coatings Technology*, **202**(9), 1681–1687 (2008).

13. W.-Y. Li, C. Zhang, X. Guo, G. Zhang, H. Liao, C. Coddet, Deposition characteristics of Al–12Si alloy coating fabricated by cold spraying with relatively large powder particles, *Applied Surface Science*, **253**(17), 7124–7130 (2007).

14. L. Ajdelsztajn, A. Zuniga, B. Jodoin, E. Lavernia, Cold-spray processing of a nanocrystalline Al–Cu–Mg–Fe–Ni alloy with Sc, *Journal of Thermal Spray Technology*, **15**(2), 184–190 (2006).

15. J. Huang, X. Yan, C. Chang, Y. Xie, W. Ma, R. Huang, R. Zhao, S. Li, M. Liu, H. Liao, Pure copper components fabricated by cold spray (CS) and selective laser melting (SLM) technology, *Surface and Coatings Technology*, **395**, 125936 (2020).

16. W. Sun, A. Bhowmik, A. W.-Y. Tan, R. Li, F. Xue, I. Marinescu, E. Liu, Improving microstructural and mechanical characteristics of cold-sprayed Inconel 718 deposits via local induction heat treatment, *Journal of Alloys and Compounds*, **797**, 1268–1279 (2019).

17. W. Sun, A. W.-Y. Tan, A. Bhowmik, I. Marinescu, X. Song, W. Zhai, F. Li, E. Liu, Deposition characteristics of cold sprayed Inconel 718 particles on Inconel 718 substrates with different surface conditions, *Materials Science and Engineering: A*, **720**, 75–84 (2018).

18. H. Aydin, M. Alomair, W. Wong, P. Vo, S. Yue, Cold sprayability of mixed commercial purity Ti plus Ti6Al4V metal powders, *Journal of Thermal Spray Technology*, **26**(3), 360–370 (2017).

19. A.W.-Y. Tan, W. Sun, A. Bhowmik, J.Y. Lek, I. Marinescu, F. Li, N.W. Khun, Z. Dong, E. Liu, Effect of coating thickness on microstructure, mechanical properties and fracture behaviour of cold sprayed Ti6Al4V coatings on Ti6Al4V substrates, *Surface and Coatings Technology*, **349**, 303–317 (2018).

20. A.W.-Y. Tan, W. Sun, A. Bhowmik, J.Y. Lek, X. Song, W. Zhai, H. Zheng, F. Li, I. Marinescu, Z. Dong, Effect of substrate surface roughness on microstructure and mechanical properties of cold-sprayed Ti6Al4V coatings on Ti6Al4V substrates, *Journal of Thermal Spray Technology*, **28**(8), 1959–1973 (2019).

21. W. Ma, Y. Xie, C. Chen, H. Fukanuma, J. Wang, Z. Ren, R. Huang, Microstructural and mechanical properties of high-performance Inconel 718 alloy by cold spraying, *Journal of Alloys and Compounds*, **792**, 456–467 (2019).

22. W. Sun, A.W.-Y. Tan, D.J.Y. King, N.W. Khun, A. Bhowmik, I. Marinescu, E. Liu, Tribological behavior of cold sprayed Inconel 718 coatings at room and elevated temperatures, *Surface and Coatings Technology*, **385**, 125386 (2020).

23. S. Yin, W. Li, B. Song, X. Yan, M. Kuang, Y. Xu, K. Wen, R. Lupoi, Deposition of FeCoNiCrMn high entropy alloy (HEA) coating via cold spraying, *Journal of Materials Science and Technology*, **35**(6), 1003–1007 (2019).

24. J.-W. Yeh, S.-K. Chen, S.-J. Lin, J.-Y. Gan, T.-S. Chin, T.-T. Shun, C.-H. Tsau, S.-Y. Chang, Nanostructured high-entropy alloys with multiple principal elements: Novel Alloy design concepts and outcomes, *Advanced Engineering Materials*, **6**(5), 299–303 (2004).

25. Y.F. Ye, Q. Wang, J. Lu, C.T. Liu, Y. Yang, High-entropy alloy: challenges and prospects, *Mater Today*, **19**(6), 349–362 (2016).
26. C. Chen, Y. Xie, L. Liu, R. Zhao, X. Jin, S. Li, R. Huang, J. Wang, H. Liao, Z. Ren, Cold spray additive manufacturing of Invar 36 alloy: microstructure, thermal expansion and mechanical properties, *Journal of Materials Science & Technology*, **72**, 39–51 (2021).
27. X. Xie, C. Chen, Z. Chen, W. Wang, S. Yin, G. Ji, H. Liao, Achieving simultaneously improved tensile strength and ductility of a nano-TiB2/AlSi10Mg composite produced by cold spray additive manufacturing, *Composites Part B: Engineering*, **202**, 108404 (2020).
28. C. Chen, Y. Xie, X. Yan, M. Ahmed, R. Lupoi, J. Wang, Z. Ren, H. Liao, S. Yin, Tribological properties of Al/diamond composites produced by cold spray additive manufacturing, *Additive Manufacturing*, **36**, 101434 (2020).
29. V. Champagne, D. Helfritch, Critical assessment 11: structural repairs by cold spray, *Materials Science and Technology*, **31**(6), 627–634 (2015).
30. R.N. Raoelison, C. Verdy, H. Liao, Cold gas dynamic spray additive manufacturing today: Deposit possibilities, technological solutions and viable applications, *Materials & Design*, **133**, 266–287 (2017).
31. V.K. Champagne Jr, O.C. Ozdemir, A. Nardi, *Practical Cold Spray*, Springer, 2021.
32. H. Renzhong, S. Wen, G. Shuangquan, X. Yingchun, L. Min, Research developments and applications of cold spray technology, *China Surface Engineering*, **33**(4), 16–25 (2020).
33. W. Sun, X. Chu, H. Lan, R. Huang, J. Huang, Y. Xie, J. Huang, G. Huang, Current implementation status of cold spray technology: A short review, *Journal of Thermal Spray Technology*, **31**(4), 848–865 (2022).
34. T. Stamey, Main circulating water pump: case study, Cold Spray Action Team (CSAT) Conference, 2016.
35. J. Frattolin, R. Barua, H. Aydin, S. Rajagopalan, L. Gottellini, R. Leask, S. Yue, D. Frost, O.F. Bertrand, R. Mongrain, Development of a novel biodegradable metallic stent based on microgalvanic effect, *Annals of Biomedical Engineering*, **44**(2), 404–418 (2016).
36. J. Frattolin, *A biomechanical evaluation of an iron and stainless steel 316L biodegradable coronary stent*, McGill University, 2018.
37. From coating to additive manufacturing - the range of cold spray applications, https://impact-innovations.com/en/applications
38. NASA additively manufactured rocket engine hardware passes cold spray, hot fire tests, https://www.nasa.gov/centers/marshall/news/releases/2021/nasa-additively-manufactured-rocket-engine-hardware-passes-cold-spray-hot-fire-tests.html
39. S. Yin, P. Cavaliere, B. Aldwell, R. Jenkins, H. Liao, W. Li, R. Lupoi, Cold spray additive manufacturing and repair: Fundamentals and applications, *Additive Manufacturing*, **21**, 628–650 (2018).
40. W. Sun, A. W.-Y. Tan, N. W. Khun, I. Marinescu, E. Liu, Effect of substrate surface condition on fatigue behavior of cold sprayed Ti6Al4V coatings, *Surface and Coatings Technology*, **320**, 452–457 (2017).
41. W. Sun, A. W.-Y. Tan, I. Marinescu, W. Q. Toh, E. Liu, Adhesion, tribological and corrosion properties of cold-sprayed CoCrMo and Ti6Al4V coatings on 6061-T651 Al alloy, *Surface and Coatings Technology*, **326**, 291–298 (2017).

42. C. Widener, M. Carter, O. Ozdemir, R. Hrabe, B. Hoiland, T. Stamey, V. Champagne, T.J. Eden, Application of high-pressure cold spray for an internal bore repair of a navy valve actuator, *Journal of Thermal Spray Technology*, 25(1), 193–201 (2016).

43. F. Gärtner, T. Stoltenhoff, T. Schmidt, H. Kreye, The cold spray process and its potential for industrial applications, *Journal of Thermal Spray Technology*, 15(2), 223–232 (2006).

44. A. Foelsche, *Nozzle Clogging Prevention and Analysis in Cold Spray*, (2020).

45. O. Stier, Fundamental cost analysis of cold spray, *Journal of Thermal Spray Technology*, 23(1), 131–139 (2014).

46. R. Nikbakht, S. Seyedein, S. Kheirandish, H. Assadi, B. Jodoin, The role of deposition sequence in cold spraying of dissimilar materials, *Surface and Coatings Technology*, 367, 75–85 (2019).

47. X. Song, J. Everaerts, W. Zhai, H. Zheng, A. W.-Y. Tan, W. Sun, F. Li, I. Marinescu, E. Liu, A. M. Korsunsky, Residual stresses in single particle splat of metal cold spray process–Numerical simulation and direct measurement, *Materials Letters*, 230, 152–156 (2018).

48. R. Ghelichi, S. Bagherifard, D. MacDonald, I. Fernandez-Pariente, B. Jodoin, M. Guagliano, Experimental and numerical study of residual stress evolution in cold spray coating, *Applied Surface Science*, 288, 26–33 (2014).

49. A. Moridi, S.M. Hassani-Gangaraj, S. Vezzú, L. Trško, M. Guagliano, Fatigue behavior of cold spray coatings: The effect of conventional and severe shot peening as pre-/post-treatment, *Surface and Coatings Technology*, 283, 247–254 (2015).

50. D. Seo, K. Ogawa, K. Sakaguchi, N. Miyamoto, Y. Tsuzuki, Parameter study influencing thermal conductivity of annealed pure copper coatings deposited by selective cold spray processes, *Surface and Coatings Technology*, 206(8–9), 2316–2324 (2012).

51. K. Spencer, M.-X. Zhang, Heat treatment of cold spray coatings to form protective intermetallic layers, *Scripta Materialia*, 61(1), 44–47 (2009).

52. W. Zhizhong, H. Chao, G. Huang, H. Bin, Cold spray micro-defects and post-treatment technologies: A review, *Rapid Prototyping Journal*, 28(2), 330–357 (2021).

53. A. Sabard, P. McNutt, H. Begg, T. Hussain, Cold spray deposition of solution heat treated, artificially aged and naturally aged Al 7075 powder, *Surface and Coatings Technology*, 385, 125367 (2020).

54. F. Gärtner, T. Stoltenhoff, J. Voyer, H. Kreye, S. Riekehr, M. Kocak, Mechanical properties of cold-sprayed and thermally sprayed copper coatings, *Surface and Coatings Technology*, 200(24), 6770–6782 (2006).

55. W. Sun, X. Chu, J. Huang, H. Lan, A. W.-Y. Tan, R. Huang, E. Liu, Solution and double aging treatments of cold sprayed Inconel 718 coatings, *Coatings*, 12(3), 347 (2022).

56. W. Sun, A. Bhowmik, A. W. Y. Tan, F. Xue, I. Marinescu, F. Li, E. Liu, Strategy of incorporating Ni-based braze alloy in cold sprayed Inconel 718 coating, *Surface and Coatings Technology*, 358, 1006–1012 (2019).

57. H. Koivuluoto, A. Coleman, K. Murray, M. Kearns, P. Vuoristo, High pressure cold sprayed (HPCS) and low pressure cold sprayed (LPCS) coatings prepared from OFHC Cu feedstock: overview from powder characteristics to coating properties, *Journal of Thermal Spray Technology*, 21(5), 1065–1075 (2012).

58. A. Sova, S. Grigoriev, A. Okunkova, I. Smurov, Potential of cold gas dynamic spray as additive manufacturing technology, *The International Journal of Advanced Manufacturing Technology*, 69(9–12), 2269–2278 (2013).

59. P. Petrovskiy, A. Sova, M. Doubenskaia, I. Smurov, Influence of hot isostatic pressing on structure and properties of titanium cold-spray deposits, *The International Journal of Advanced Manufacturing Technology*, 102(1), 819–827 (2019).

60. H. Wu, X. Xie, M. Liu, C. Verdy, Y. Zhang, H. Liao, S. Deng, Stable layer-building strategy to enhance cold-spray-based additive manufacturing, *Additive Manufacturing*, 35, 101356 (2020).

61. H. Tabbara, S. Gu, D. McCartney, T. Price, P. Shipway, Study on process optimization of cold gas spraying, *Journal of Thermal Spray Technology*, 20(3), 608–620 (2011).

62. C. Chen, Y. Xie, C. Verdy, H. Liao, S. Deng, Modelling of coating thickness distribution and its application in offline programming software, *Surface and Coatings Technology*, 318, 315–325 (2017).

63. D. Vanerio, J. Kondas, M. Guagliano, S. Bagherifard, 3D modelling of the deposit profile in cold spray additive manufacturing, *Journal of Manufacturing Processes*, 67, 521–534 (2021).

64. Z. Cai, *Programmation robotique en utilisant la méthode de maillage et la simulation thermique du procédé de la projection thermique*, Université de Technologie de Belfort-Montbéliard, 2014.

65. R. Gadow, A. Candel, M. Floristán, Optimized robot trajectory generation for thermal spraying operations and high quality coatings on free-form surfaces, *Surface and Coatings Technology*, 205(4), 1074–1079 (2010).

66. D.H.L. Seng, Z. Zhang, Z.-Q. Zhang, T.L. Meng, S.L. Teo, B.H. Tan, L. Qizhong, P. Jisheng, Influence of spray angle in cold spray deposition of Ti-6Al-4V coatings on Al6061-T6 substrates, *Surface and Coatings Technology*, 432, 128068 (2022).

67. J. Pattison, S. Celotto, R. Morgan, M. Bray, W. O'Neill, Cold gas dynamic manufacturing: A non-thermal approach to freeform fabrication, *International Journal of Machine Tools and Manufacture*, 47(3–4), 627–634 (2007).

68. D. Kotoban, S. Grigoriev, A. Okunkova, A. Sova, Influence of a shape of single track on deposition efficiency of 316L stainless steel powder in cold spray, *Surface and Coatings Technology*, 309, 951–958 (2017).

69. Z. Djurić, P. Grant, An inverse problem in modelling liquid metal spraying, *Applied Mathematical Modelling*, 27(5), 379–396 (2003).

70. S. Duncan, P. Jones, P. Wellstead, A frequency-domain approach to determining the path separation for spray coating, *IEEE Transactions on Automation Science and Engineering*, 2(3), 233–239 (2005).

71. H. Koivuluoto, J. Larjo, D. Marini, G. Pulci, F. Marra, Cold-sprayed Al6061 coatings: Online spray monitoring and influence of process parameters on coating properties, *Coatings*, 10(4), 348 (2020).

72. I.M. Nault, G.D. Ferguson, A.T. Nardi, Multi-axis tool path optimization and deposition modeling for cold spray additive manufacturing, *Additive Manufacturing*, 38, 101779 (2021).

73. Z. Cheng, H. Wang, G.-R. Liu, Deep convolutional neural network aided optimization for cold spray 3D simulation based on molecular dynamics, *Journal of Intelligent Manufacturing*, 32(4), 1009–1023 (2021).

74. D. Ikeuchi, A. Vargas-Uscategui, X. Wu, P.C. King, Neural network modelling of track profile in cold spray additive manufacturing, *Materials*, **12**(17), 2827 (2019).
75. D. Ikeuchi, A. Vargas-Uscategui, X. Wu, P.C. King, Data-efficient neural network for track profile modelling in cold spray additive manufacturing, *Applied Sciences*, **11**(4), 1654 (2021).

Chapter 5

Progress in solid-state additive manufacturing of composites

Bhavesh Chaudhary, Mahesh Patel, Neelesh Kumar Jain and Jayaprakash Murugesan

Indian Institute of Technology Indore, Indore, India

Vivek Patel

University West, Trollhättan, Sweden

5.1 INTRODUCTION

Composites are a new class of engineering materials that are made up of two or more distinct constituents that are mixed using an appropriate manufacturing procedure. These constituents are distinctly separated at the interface boundaries and are known as matrix and reinforcing particles. Matrix materials includes metals, polymers, or ceramics whereas reinforcement also can be of these three types [1]. Composites are principally divided into four types as follows: (i) *polymer matrix composites*, in which a polymer such as epoxy is used as the matrix material that is reinforced with very fine diameter fibers such as carbon, glass, etc. The reinforcing fibers can be either continuous or discontinuous. Continuous fibers can be used in forms such as strands, roving, fabric, etc. Discontinuous fibers, by contrast, can be particulate, whiskers, or flakes, (ii) *metal matrix composites (MMC)*, in which a metal or an alloy is the continuous phase and reinforcements are embedded in them. Reinforcements can be found in the form of particulates, short fibers, flakes etc. These composites possesses high transverse strength and modulus, high shear strength and modulus, high service temperature, low thermal expansion, very low moisture absorption, dimensional stability, high electrical and thermal conductivities, better fatigue and damage resistance, ease of joining and resistance to most radiations including ultraviolet (UV) radiation etc., (iii) *ceramic matrix composites*, in which either continuous or discontinuous reinforcements are embedded in a monolithic ceramic material. These composites are suitable for applications where high mechanical properties are desired at high service temperatures, and (iv) *carbon–carbon composites*, in which carbon fiber reinforcements are embedded in a carbon matrix. Their high tensile and compressive strengths, along with their ability to maintain high fatigue strength at elevated temperatures, make them suitable for a variety of high-end applications in aerospace and other industries. Examples include brake disks for aircraft, nose cones for reentry vehicles, and nozzle

DOI: 10.1201/9781032616025-5

throats [2]. Our emphasis in this chapter is principally on the MMCs. Thus, unless otherwise specifically stated, composites in the subsequent text would mean MMCs.

Metal matrix composites (MMCs) perform better than polymeric matrix composites in applications that demand long-term durability under harsh conditions, particularly at high temperatures. Metals generally possess higher yield strength and modulus than polymers, making them well-suited for applications requiring strong transverse strength and modulus, as well as compressive strength for the composite. Additionally, metals can be plastically deformed and strengthened through a range of thermal and mechanical treatments, providing further advantages. However, metals have a number of drawbacks, including high densities, high melting points, and a tendency for corrosion at the interface between the matrix and the reinforcement [3]. Aluminum (Al) and titanium (Ti) are the two most frequently used metal matrices. Magnesium, while lightweight, has a high affinity for oxygen, leading to atmospheric corrosion and limiting its use in numerous applications. On the other hand, beryllium has the highest tensile modulus of all structural metals and is also lightweight. However, its extreme brittleness makes it unsuitable as a matrix material. Superalloys, comprising nickel and cobalt, have also been utilized as matrices, but the alloying elements in these materials tend to amplify fiber oxidation at high temperatures [4]. As a result, aluminum and its alloys have received more attention as matrix materials in MMCs. Pure Al has been utilized commercially because of its high corrosion resistance. Aluminum alloys with high tensile strength-to-weight ratios, such as 1100, 201, 6061, are widely used. Carbon fibers are used as reinforcement materials with an Al matrix, but they interact with aluminium to generate aluminium carbide (Al_4C_3) at manufacturing temperatures of 500°C or higher, which drastically affects the mechanical characteristics of the composite. To prevent fiber deterioration and improve wetting with the Al alloy matrix, carbon fibers have been coated with protective layers of titanium boride (TiB_2) or sodium. However, carbon fiber-reinforced Al composites are highly susceptible to galvanic corrosion, as the carbon fibers act as a cathode with a corrosion potential 1V higher than that of aluminum. SiC is a more prevalent reinforcement for Al alloys [2]. The most effective titanium alloys in MMCs include α, β alloys (e.g., Ti-6Al-9V) and metastable β-alloys (e.g., Ti-10V-2Fe-3Al). Compared with Al alloys, these Ti alloys exhibit higher strength-to-weight ratios and improved strength retention at 400°C–500°C. Ti alloys have a lower thermal expansion coefficient than reinforcing fibers, reducing the thermal mismatch between them. One significant challenge with Ti alloys during normal manufacturing temperatures is their high reactivity with boron and Al_2O_3 fibers. However, Borsic (boron fibers coated with silicon carbide) and silicon carbide (SiC) fibers exhibit lower reactivity with titanium. Additionally, coating boron and SiC fibers with carbon-rich layers has been shown to improve the retention of tensile strength [4].

Composites are conventionally manufactured by mixing micro- or nano-sized reinforcing particles in a base matrix using different processes such as laser, electron beam irradiation, spray deposition, powder metallurgy, casting, injection molding, diffusion bonding, and so on. These conventional processes manufacture composites in liquid phase at extreme temperatures, resulting in the production of intermetallic compounds and unwanted phases between the substrate and reinforcement. These drawbacks necessitate the manufacturing of composites at temperatures lower than the melting points of the matrix [5]. Additive manufacturing (AM), sometimes known as 3D printing, is a novel process that is used to create high-quality, lightweight composites for a variety of industrial applications. It provides a platform for constructing features by adding material layer by layer using modelling tools such as computer-aided design (CAD) or three-dimensional (3D) object scanning cameras. The resulting composites are highly precise with regard to the final dimensions, usually negating the need for post-deposition machining. However, some composites may still require post-deposition machining to achieve the desired properties and strength. AM allows for design freedom and provides a means for manufacturing components in a single step. Consequently, AM processes can produce complex-shaped composites with less material than conventional machining processes [6]. Metal AM processes are classified into two types depending on the melting requirements of the feedstock material and the kind of heat source used, as illustrated in Figure 5.1. These are *fusion-based AM (FBAM)* and *solid-state AM (SSAM)* processes. FBAM processes, as the name indicates, deal with complicated 3D structures by melting their feedstock material in a confined space. In contrast, in the SSAM processes, the feedstock material is not melted; instead, 3D composites are built up using other mechanisms, such as friction, pressure, and velocity [7]. The FBAM processes can be further classified into the following two subcategories: (a) Beam-based processes, namely electron-beam based processes (i.e., electron beam melting [8]), and laser-beam based processes such as selective laser melting [9], selective laser sintering [10], laser-engineered net shaping [11], direct metal laser sintering [12], and (b) arc-based processes: processes which use the arc for material deposition generally in wire form and referred to as wire arc additive manufacturing (WAAM), i.e., gas metal arc or gas tungsten arc [13], and the processes which use the arc for plasma formation such as plasma transferred arc and micro-plasma transferred arc [14] and then the plasma/micro-plasma is used as the heat source. The SSAM processes are deformation-based processes which are characterized by the consolidation of the feedstock material through plastic deformation. It can be induced by (i) ultrasonic vibrations between two foils in *ultrasonic additive manufacturing* [15], (ii) the supersonic impact of powder particles on the substrate or the previously deposited layer in *cold spray additive manufacturing* [16], or (iii) friction stirring in the processes such as *friction stir additive manufacturing* and *additive friction stir deposition* [6]. In SSAM processes, heat is generated as a byproduct, and the temperature

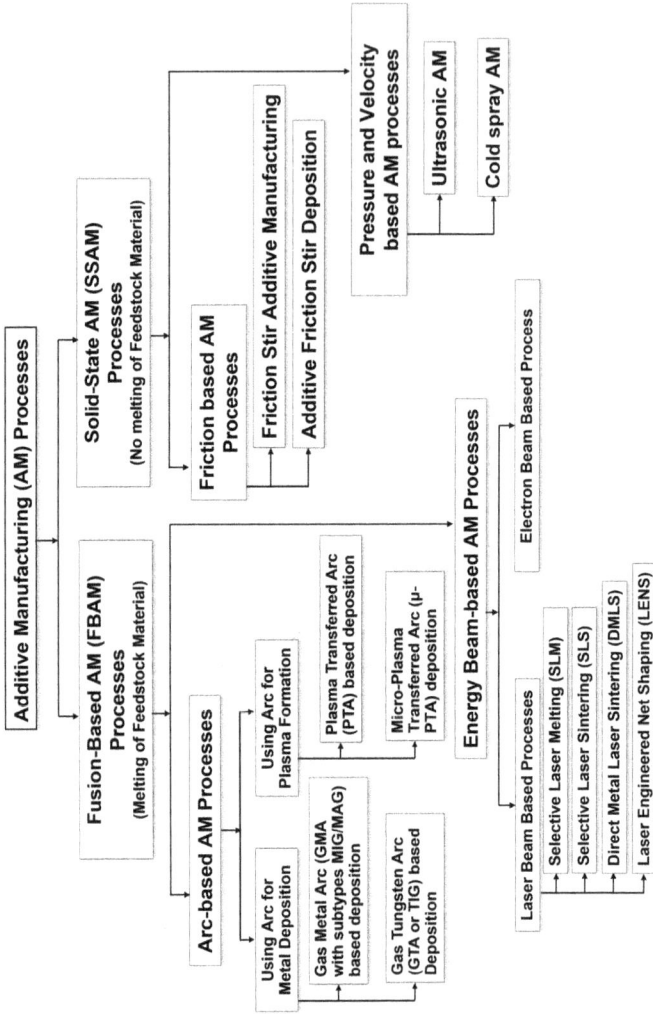

Figure 5.1 Classification of additive manufacturing processes based on the type of heat sources.

rise is influenced by the physical and thermal properties of the feedstock and substrate materials. The microstructure of components produced using SSAM processes is typically characterized by recrystallized fine grains in regions where severe plastic deformation occurs [7].

The SSAM procedures are deformation-based methods that include the aggregation of feedstock material by plastic deformation. It can be caused by (i) ultrasonic vibrations among two foils in ultrasonic additive manufacturing [15], (ii) the supersonic impact of powder particles on the substrate or previously deposited layer in cold spray additive manufacturing (Pathak and Saha, 2020), or (iii) friction stirring in processes like friction stir additive manufacturing and additive friction stir deposition [6]. The SSAM processes generate heat as a byproduct, and the magnitude of the temperature rise is determined mostly by physical as well as thermal characteristics of the feedstock and substrate materials. In regions of extreme plastic deformation, the microstructure of components made by the SSAM procedures is characterized by recrystallized small grains.

The use of FBAM processes to additively manufacture composites using Al alloy as a matrix material has been explored over the past few years. But these processes are found to suffer from the following limitations: (i) a higher thermal gradient which makes it challenging to manufacture components from the difficult to additively manufacture Al alloys of 2xxx, 6xxx, and 7xxx series [17], (ii) the presence of solidification-related defects such as porosities and hot cracking, (iii) columnar grains and cast microstructure along the build direction yielding anisotropic mechanical properties in most of the cases [18], (iv) the supply of excessive heat leading to a spattering of powder particles deteriorating deposition quality whereas the supply of insufficient heat results in both an inter-layer and an intra-layer lack of fusion, and (v) the frequent heating of the already deposited layers during the deposition of a layer leads to the formation of undesirable phases [19]. These limitations have necessitated development of SSAM processes to deposit a feedstock material without melting to overcome the limitations of FBAM processes [20].

The absence of melting of the feedstock material in SSAM processes allows for easy control of high temperature gradients and eliminates the solidification-related defects which are usually observed in FBAM processes [17]. Moreover, SSAM processes have several distinct benefits, such as reduced oxidation, equiaxed grains, no residual stress-induced cracking, no requirement for a protective gas environment, no hazardous fume production, ease in manufacturing composites, and no component size constraints [21]. SSAM processes have a wide range of applications and broadly accepted as the processes of future with enormous potential. There is a strong need for review on the manufacturing of composites by SSAM processes to understand the widespread applicability of SSAM in an improved material manufacturing sector. This chapter presents a brief review on manufacturing of composites by SSAM processes. It has been compiled taking into account relevant SSAM processes and different aspects in a logical sequence. Initially, introduction to composites and various manufacturing processes, including AM, is highlighted. This is followed by an explanation of various SSAM processes with their working principle, methods for their feedstock preparation, deposition geometry and reinforcement volume, microstructure and mechanical properties. Finally, a comprehensive review of the challenges and perspectives of SSAM in the manufacturing of composites have been discussed. Towards the end of this chapter, future directions pertaining to composites are presented.

5.2 MANUFACTURING OF COMPOSITES BY DIFFERENT SSAM PROCESSES

The following section describes the working principle of different SSAM processes, different methods for preparation of their feedstock material for composite manufacturing, and deposition geometry, microstructure, and mechanical properties of the composites.

5.2.1 Ultrasonic additive manufacturing process

The following sub-sections describe different aspects of ultrasonic additive manufacturing (UAM) process.

5.2.1.1 Working principle

The UAM process involves the use of foils with thicknesses in fractions of a millimeter as feedstock material. These foils are bonded together using high-frequency ultrasonic vibrations (typically at 20 kHz) and applying a normal force, as depicted in Figure 5.2. The ultrasonic vibrations are produced by a low (1–2 kW) or high (up to 9 kW) power transducer and are transmitted

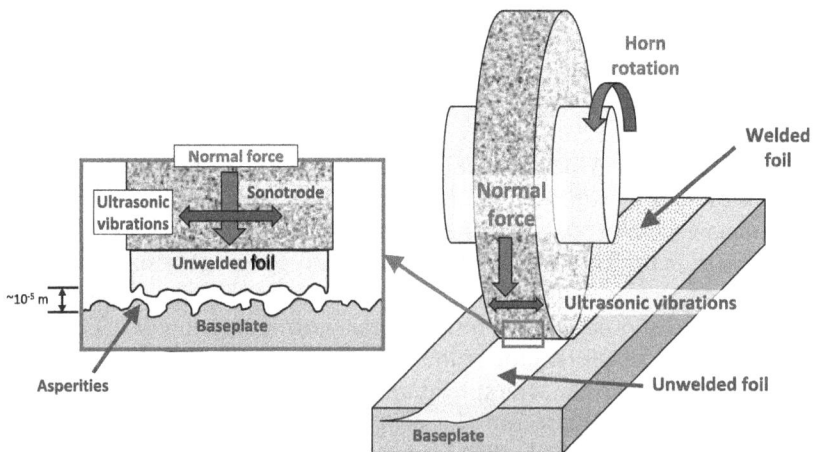

Figure 5.2 Schematic diagram depicting working principle of ultrasonic additive manufacturing process [22].

to the foil surface via a sonotrode that oscillates parallel to the build direction and along the length of the foil. Before the placement of the next foil, the rough surface of the sonotrode leaves microasperities on the foil surface. High-frequency vibrations and normal force cause severe plastic deformation between two adjacent foils, with strain rates as high as 10^4 to 10^5 s^{-1}. Bonding is accomplished by the intense plastic deformation at the interface and the collapse of the micro-asperities [7]. The desired shape is obtained by removing the excess material.

Table 5.1 presents different MMCs manufactured by the UAM process and subsequently major findings from these MMCs has been briefly explained.

UAM has a wide range of applications, including the manufacture of soft materials like Al alloys, as well as harder materials such as copper, nickel, steel, and titanium. UAM is also capable of producing dissimilar materials, such as Al-Ti, Al-Cu, Al-steel, Al-NiTi, and Ni-steel, as well as metal matrix composites embedded with elements like SiC fibers, shape memory alloy fibers, dielectric materials (used in printed circuit boards), electronic circuitry, glass fibers, fiber optic sensors, and carbon fibers [27]. Guo et al. [22] utilized the UAM process to produce Al 6061/CFRP MMCs consisting of 12 and 15 layers, as illustrated in Figure 5.3. The tensile tests conducted on the samples revealed peak loads of 4677 N and 7238 N for the 12-layered sample with a thickness of 1.8 mm and the 15-layered sample with a thickness of 2.2 mm, respectively. The failure mechanisms of both samples were

Table 5.1 Metal matrix composites manufactured by ultrasonic additive manufacturing process

Feedstock material	Process parameters and number of deposition layers	Ref.
Al 6061/Carbon fiber reinforced polymer (CFRP)	No. of layers: Total 12–15 layers of Al 6061 and CFRP.	[22]
13% NiTi as reinforcement fibers filled inside the holes made in Al 6061-T6 matrix sheets	Ultrasonic amplitude: 34.6 μm Applied normal force: 6000N Travel speed of sonotrode: 5076 mm/min Dimensions of MMC: 5.49 × 74 × 1.27 mm	[23]
Al 3003 H18/Ti6Al 4V (Embedding of electronic circuitry)	Ultrasonic amplitude: 25 μm Applied normal force: 1600N Travel speed of sonotrode: 1200 mm/min No of layers: 1st two layers-Al 3003, 3rd layer-Ti6Al4V and last layer of Al 3003	[24]
Al 6061/YSZ	Ultrasonic amplitude: 30 μm Applied normal force: 4000N Travel speed of sonotrode: 2032 mm/min No. of layers: 1st layer-Al 6061, 2nd layer-YSZ and next 21 layers of Al 6061 foil	[25]
Al 3003/NiTi	No. of layers: 1st two layers-Al 3003, 2nd layer-NiTi and next two layers of Al 3003	[26]

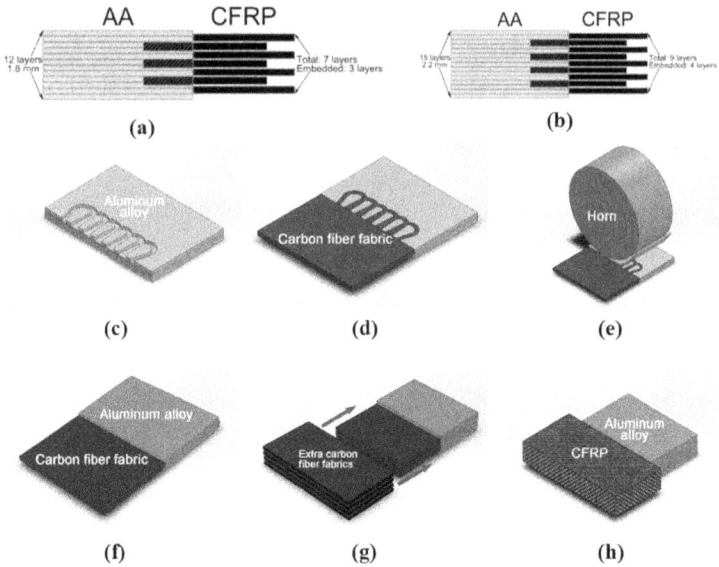

Figure 5.3 Schematic cross-section of (a) 12-layered deposition and (b) 15-layered deposition; Manufacturing process of Al 6061/CFRP MMC showing (c) AA matrix is built with UAM and channels are machined, (d) dry CF tow loops are placed in the channels, (e) one layer of AA foil is deposited by the UAM horn, (f) one layer of CF tow loops is embedded; (g) three layers of CF tows are embedded by repeating steps (c) to (f), extra CF fabrics are inserted to match the thickness of CFRP to AA, (h) CFRP is cured with epoxy to complete the CFRP-AA manufacturing (*AA-aluminum alloy and CF-Carbon fiber) [22].

analyzed and suggested that the mode of failure could be controlled by adjusting the ratio of carbon fiber to aluminum alloy (AA). In their research, Chen et al. [23] employed the UAM process to produce an Al 6061/NiTi MMC. The placement and encapsulation of the NiTi fibers were facilitated by creating a deep channel with a size of 0.356 mm in the Al matrix, using a ball-nose end mill with a diameter of 0.397 mm. The findings indicated that heating the MMC caused a decrease in the coefficient of thermal expansion (CTE), from 20×10^{-6}/K at temperatures below 40°C to 11×10^{-6}/K between 40°C and 100°C. The reduction in CTE at higher temperatures was attributed to a phase transformation from martensite to austenite in the NiTi shape memory alloy wires, which caused axial contraction and opposed thermal expansion. Li et al. [24] used the UAM process to create a novel multifunctional MMC composed of Al 3003 H18 as the matrix material, with printed electrical circuitries embedded directly within the interlaminar region. The manufacturing process involved three stages. First, a substrate consisting of the first two layers of Al 3003 and a layer of Ti6Al4V was created using UAM (Figure 5.4a). Next, the electrical circuitry was printed

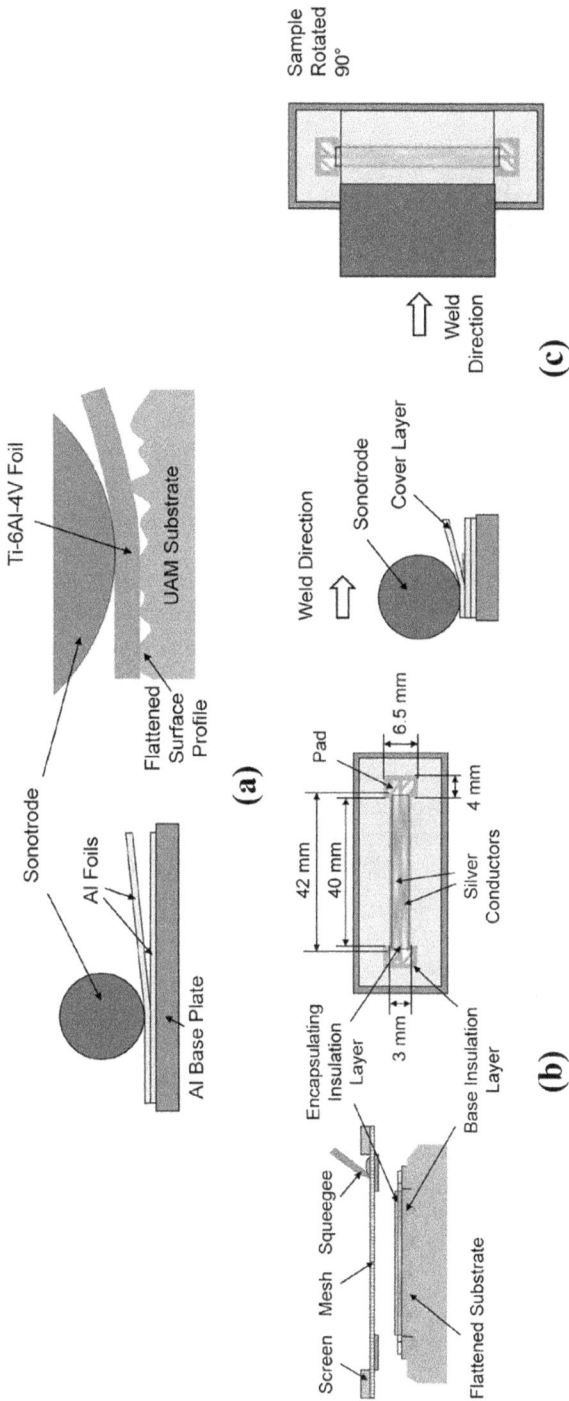

Figure 5.4 Manufacturing of substrate (Al 3003/Ti6Al4V) which includes (a) first two layers of Al 3003 and next layer of Ti6Al4V by UAM; (b) Electrical circuitry printing using screen printing process; and (c) Covering of electronic circuitry (consists of 520 series insulators, and P58 and 4D3 conductive traces) using fabrication of Al 3003 sheet on the substate by UAM [24].

using screen printing (Figure 5.4b). Finally, the circuitry was covered using the fabrication of an Al 3003 sheet on the substrate by UAM (Figure 5.4c). Remarkably, the printed electrical structures remained functional even after being encapsulated at higher UAM energies without the need for extra pockets or auxiliary materials to shield them from the embedding process. Additionally, a surface-flattening procedure was created to minimize the thickness of the printed circuitry, enhance the integrity of the MMCs, and increase their mechanical strength in order to completely prevent the possibility of short-circuiting between the metal matrices and printed conductors. The conductivity of both silver inks was maintained after UAM embedding, even under relatively high UAM process energy, using the robust 520 Series insulation shell for protection. The P58 flake-filled silver ink demonstrated significantly better conductivity after UAM embedding than its 4D3 particle-filled counterpart, as described by their relative resistivity. The P58 ink had a resistivity of 6.6×10^{-4} Ω cm at low embedding energies and 1.2×10^{-3} Ω cm at high embedding energies, whereas the resistivity of the 4D3 ink measured almost twice these values at 9.6×10^{-4} Ω cm at lower energies and 2.2×10^{-3} Ω cm at high embedding energies, making the P58 ink more desirable for practical applications. Deng et al. [25] manufactured Al 6061/YSZ (yttria-stabilized zirconia) MMC using the UAM process. Its schematic is shown in Figure 5.5. The shear strengths of the as-welded and heat-treated MMCs are 72 MPa and 103 MPa, respectively. Hahnlen et al. [26] manufactured Al 3003/NiTi MMC using the UAM process, as depicted in Figure 5.6. The objective was to investigate the bonding mechanism between NiTi and Al and to determine the interface failure. By utilizing a constitutive composite model, the average shear strength of the interface was calculated to be 7.28 MPa. The findings indicated that the shear stress at the interface is inversely proportional to the length of the NiTi foil. Therefore, increasing

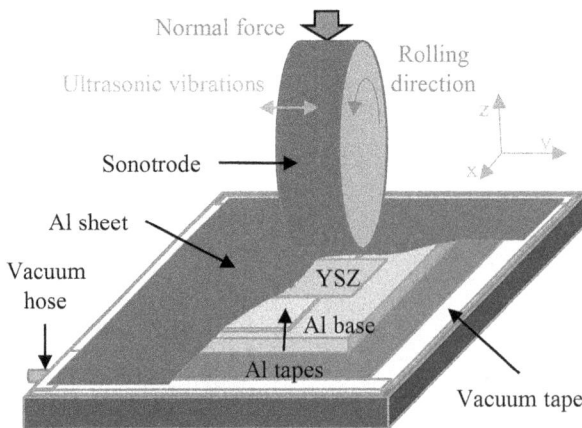

Figure 5.5 Schematic of joining YSZ films on aluminum using UAM system [25].

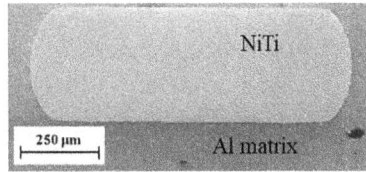

Figure 5.6 Cross-section of an UAM-manufactured Al/NiTi composite with 254 μm by 762 μm NiTi ribbon [26].

the length of the composite results in a considerable increase in the interface failure temperature, which prevents irreversible damage to the MMC.

5.2.1.2 Feedstock material and its preparation

The matrix material used in the UAM process is normally in the form of foil, whereas the reinforcement can be in form of foil, powder, or flakes. Foils can be deposited either directly, layer by layer, or indirectly, where two foils are deposited together, when connecting a soft foil to a harder foil using UAM. For indirect deposition, two separate methods are employed: (i) layering the harder layer on top of the harder layer, and (ii) layering the harder layer on top of the soft layer. It was discovered that the latter method resulted in improved bonding density and joint quality at the interfaces [28]. Soft materials are easier to bond with fewer voids using UAM. Figure 5.7 displays cross-sections of Al parts produced with multiple layers of Al 3003 using two different heat treatments. The part manufactured using the harder grade, Al 3003 H18, had considerably more voids at the interfaces than the one made using the softer grade, Al 3003 O. To reduce the number of voids during the bonding of harder materials, more energy may be necessary, along with potentially higher pressure and oscillation amplitude. When

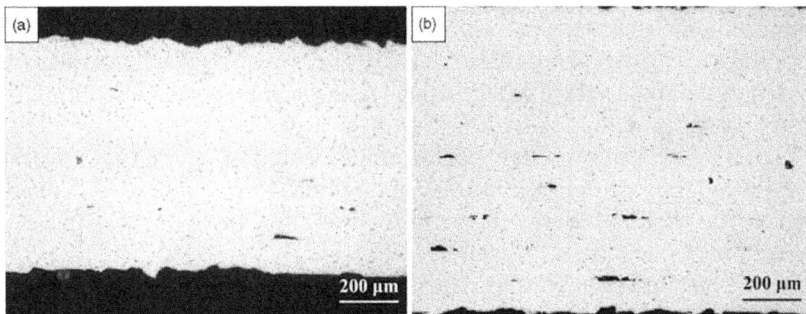

Figure 5.7 Cross-section showing microstructures of Al alloy foils manufactured by ultrasonic additive manufacturing process: (a) 3003 O, 8 layers and (b) 3003 H18, 10 layers [29].

reinforcement is in the form of powders or flakes, channels or bores are created in the substrate sheet, and then a foil made of the same material as the substrate or a different material is placed over them and closed using UAM [22], as depicted in Figure 5.3.

5.2.1.3 Microstructure and mechanical properties

Temperature generation during AM directly affects microstructure of depositions. It depends on the type and thermal properties of the feedstock material and substrate which generally lies between 0.3 and 0.5 T_m for the UAM process (where T_m is melting point of the feedstock material) [22]. It was found to increase with the increase in vibration amplitude and shear strength of feedstock material (for example, Cu heats up more than Al) in the UAM process [30]. It has been discovered that under this process internal heat generation caused by plastic deformation prevents the material from strain hardening, but localized strain results in void formation which causes brittle cracking at the interfaces between the deposition layers [31]. Elongated grains with a certain rolling texture are typically present in the feedstock material (i.e., foils) used in the UAM. Therefore, parts manufactured by them possess non-homogenous microstructure consisting of elongated grains in the bonded foils/sheets and recrystallized fine grains at their interface [32]. The UAM process has been found to be very useful in manufacturing embedded electrical circuits. The microstructure of embedded 4D3 and P58 conductors between Al 3003 H18 was found in good condition, i.e. there is an absence of obvious deformation and cracks as shown in Figure 5.8.

Porosity in the additively manufactured part is a critical factor when considering its mechanical properties. The SSAM processes mechanically disrupt the oxides on the feedstock material surface, meaning therefore that they are present in the manufactured parts. The presence of a continuous oxide film or circular oxide clusters is found at the interface in the parts manufactured by the UAM process. It typically gives 1–2% porosity with optimized process parameters and has been shown to be capable of improvement through post-AM heat treatment. Al/YSZ MMC showed an absence of voids at the bulk interface when a single layer of YSZ film is built within the Al matrix as shown in Figure 5.9.

The elongation of multi-layer depositions created using the UAM process is less than 50% of the reference (i.e., compared to wrought alloys) due to voids at the interface between the foils during the UAM process [31]. An important obstacle for UAM components is anisotropy. In particular, the degree of metallurgical bonding between successive deposition layers affects both the strength and the ductility, which vary significantly along and perpendicular to the build direction. Significant decreases in tensile strength and brittle fracture have been recorded for UAM-made composites along the build direction compared to the deposition direction [31]. It is due to the

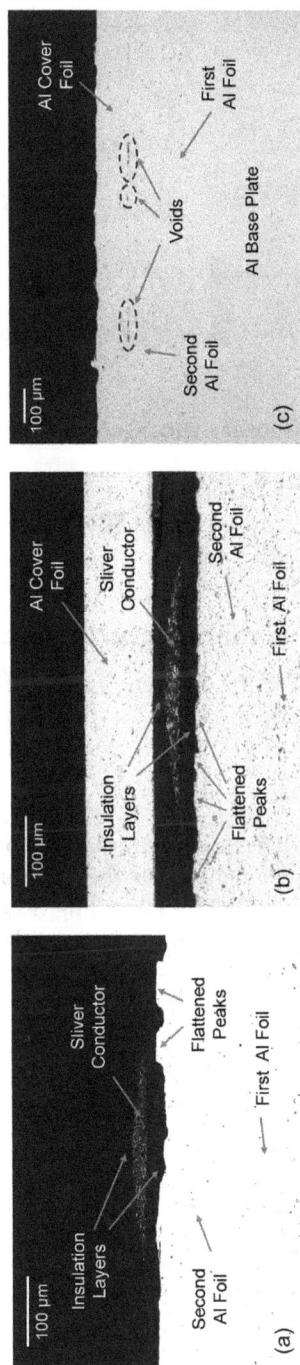

Figure 5.8 (a) cross-section of printed circuitry before embedding; (b) cross-section of circuitry embedding area; (c) cross-section of Al-Al interface [24].

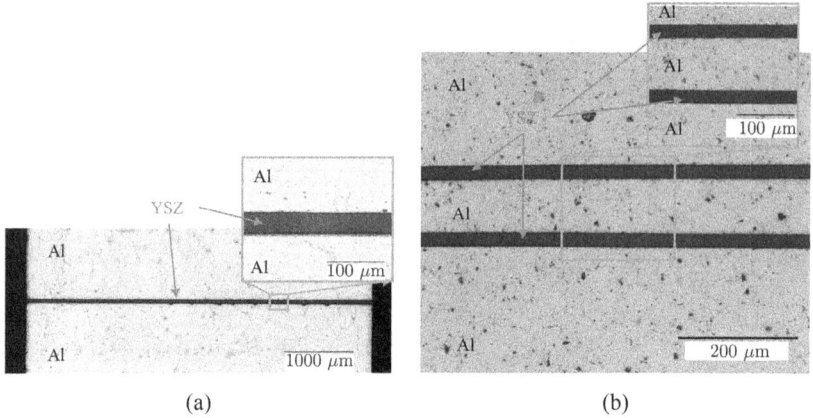

Figure 5.9 Optical micrograph showing two layers of YSZ films welded within an Al structure [25].

presence of oxide layer between two consecutive foils [33] and void coalescence due to localized strain at the interface [34].

5.2.2 Cold spray additive manufacturing process

The following sub-sections describe different aspects of the cold-spray additive manufacturing (CSAM) process.

5.2.2.1 Working principle

The CSAM process employs feedstock material in powder form and uses heated compressed air/nitrogen/helium as the propulsive gas to accelerate the feedstock material to a velocity exceeding 300 m/s, which impacts the substrate, as illustrated in Figure 5.10. This results in localized plastic

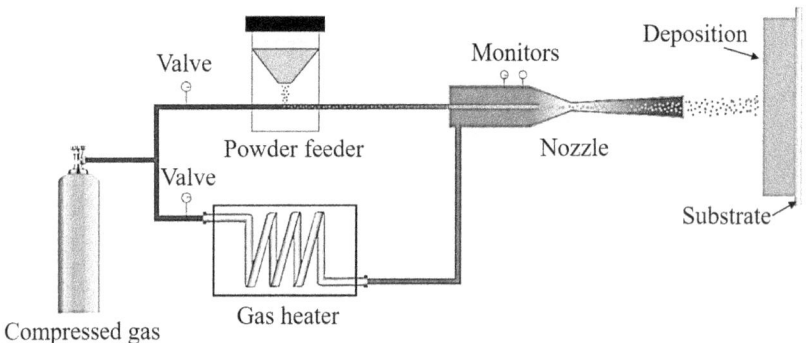

Figure 5.10 Schematic diagram of the cold spray additive manufacturing process [35].

deformation at the inter-particle and particle–substrate interfaces, inducing deposition through local metallurgical bonding and mechanical interlocking. The heating temperature in the CSAM process must be carefully controlled to ensure that the feedstock material remains in a solid-state throughout the entire deposition process [35].

Table 5.2 presents different MMCs manufactured by the CSAM process and subsequently major findings from these MMCs has been briefly explained.

Various MMCs have been additively manufactured by the CSAM process. Typical Al-based MMCs that can be found in the literature include Al/SiC, Al/TiB2, Al/B_4C, Al/Al_2O_3, Al/TiN, Al/CNT etc. Khomutov et al. [36] employed the CSAM process to manufacture a multi-track, 20-layered Al 7075-SiC MMC measuring $50 \times 60 \times 10$ mm and conducted hot-isostatic pressing (HIP) as a post-deposition heat treatment. It was determined that the maximum percentage volume of SiC reinforcement should not exceed 21–24% to achieve a maximum deposition efficiency of 50%. The SiC inclusions were uniformly distributed at the AA7075 splat boundaries, and partial fragmentation was observed. The HIP process was shown to improve the contact between splat boundaries and also facilitated the redistribution of hardening phases. However, neither the T6 heat treatment nor the HIP showed a significant improvement in ductility compared to as-built samples. Xie et al. [37] utilized the CSAM process to manufacture AlSi10Mg/TiB_2 MMC, which was subsequently processed using friction stir processing (FSP) method, as depicted in Figure 5.11. The MMC exhibited a compact structure with some TiB_2 clusters dispersed within the Al matrix. While the as-sprayed pure AlSi10Mg and AlSi10Mg/TiB_2 MMC possessed high tensile strength, they had almost no ductility owing to limited metallurgical bonding and a work-hardening effect. FSP treatment improved the tensile strength and ductility of the AlSi10Mg/TiB_2 MMC by redistributing and refining the TiB_2 nanoparticles and enhancing interfacial bonding through in-situ chemical reaction during gas atomization. Klinkov et al. [38] manufactured Al/B_4C MMC using the CSAM process. It was found that the highest productivity of the spraying process is achieved when using the optimal powder composition with a volume content of boron carbide of about 0.5. Qiu et al. [39] utilized the CSAM process to manufacture a 5-mm-tall Al A380/Al_2O_3 MMC. The study found that the deposition of spherical Al_2O_3 particles resulted in a severe tamping effect, leading to reduced porosity ($\sim$0.65%), low Al_2O_3 deposition efficiency, refined average grain size ($<$0.7 µm), and a 70% increase in strength ($\sim$390 MPa) compared to pure A380 alloy ($\sim$230 MPa). However, irregularly shaped Al_2O_3 particles exhibited a weak tamping effect, resulting in slightly lower porosity ($\sim$1.15%) and larger average grain size ($>$0.7 µm) but still with a 40% increase in strength ($\sim$330 MPa) compared to pure A380 alloy. The fragmentation of Al_2O_3 particles and the resulting large number of weakly bonded interfaces in A380/Al_2O_3 MMC contributed to this effect.

Table 5.2 Metal matrix composites manufactured by the cold spray additive manufacturing process

Feedstock material	Process parameters and type of deposition	Strength (MPa)	Microhardness (HV)	Ref.
Al 7075/20-22 vol.% SiC powder	Gas: Nitrogen (35 bar) Temperature: 550°C Nozzle speed: 3000 mm/min Nozzle displacement: 3 mm No. of layers: 20	Compressive strength As-sprayed: 720 ± 22 After HIP: 640 ± 13	As-sprayed: 156 ± 5 After HIP: 123 ± 4.1	[36]
AlSi10Mg/TiB$_2$ powder	Gas: Helium (1.8 MPa) Temperature: 320°C Nozzle speed: 6000 mm/min No. of layers: 20	Tensile strength As-sprayed: ~350 MPa with less than 1% elongation) After FSP: 365 ± 35 MPa with 16% elongation)	As-sprayed:179 ± 5 After FSP: 110 ± 3	[37]
Al/B$_4$C powder	Gas: Air (4 MPa) Temperature: 573°C Nozzle speed: 6000 mm/min Nozzle displacement: 3 mm No. of layers: 1	-	-	[38]
A380/ Al$_2$O$_3$	Gas: Air (4 MPa) Temperature: 500°C Nozzle speed: 6000 mm/min Powder feed rate: 25 ± 5 g/mm Standoff distance: 30 mm Total height of MMC: 5 mm	Tensile strength 390 MPa	-	[39]
Al /B$_4$C	Gas: Air (2.2 MPa) Temperature: 300–350°C Nozzle speed: 3 mm/min Powder feed rate: 15 ± 5 g/mm Standoff distance: 15 mm Total height of MMC: 6 mm	Tensile strength As sprayed: 38; HT@200°C: 44 HT@300°C: 56 HT@400°C: ~58 HT@500°C: ~60	As sprayed: 58 ± 2.8 HT@200°C:56±2.3 HT@300°C: 42 ± 2.6 HT@400°C: 38 ± 1.5 HT@500°C: 35 ± 1.3	[40]

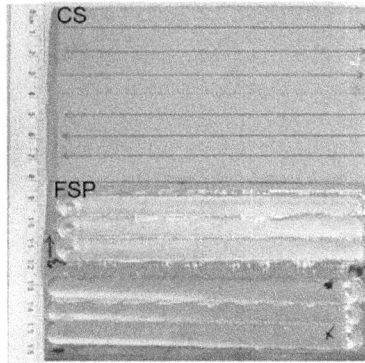

Figure 5.11 Photographs of the as-sprayed TiB$_2$/AlSi10Mg composite parts before and after FSP treatment [37].

Heat treatment at 350°C for 4 hours in a muffle furnace resulted in relatively low strength but significantly improved ductility in all MMCs by minimizing poorly bonded splats and converging pores to form larger-sized pores. Tariq et al. [40] utilized the CSAM process to fabricate a 6-mm-thick Al/B$_4$C MMC for neutron-shielding applications, as depicted in Figure 5.12. Microstructure examination confirmed that the MMC did not contain any cracks or undesired phases. The MMC was found to have entirely brittle behavior in the as-sprayed state as well as after heat treatment at 200°C, which was attributed to the presence of a large number of intersplat defects along with intensive cold working. The MMC heat-treated at 500°C exhibited the highest ductility (1.4%) and strength (60 MPa) with a minimum porosity level of 1.9%. Heat treatment did not affect neutron-shielding performance, as approximately 50–55% attenuation of the thermal neutron was achieved in the as-sprayed and heat-treated MMC samples with a thickness of 5 mm.

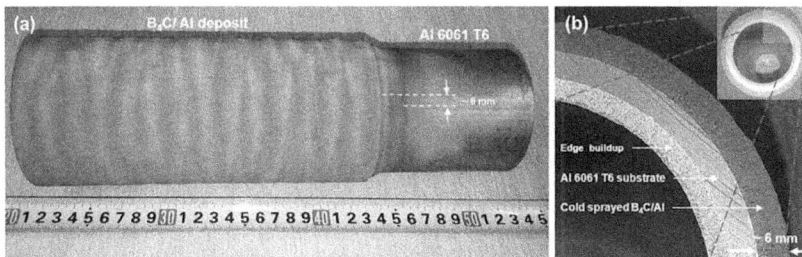

Figure 5.12 Digital images of cold-sprayed B4C/Al bulk coating showing (a) lateral view and (b) cross sectional view [40].

5.2.2.2 Feedstock preparation

It has been widely accepted that the nature of feedstock powders directly affects the microstructure and properties of CS-processed deposits or components. Therefore, it is important to choose a suitable powder preparation process. The Following are important processes for preparation of composite powders:

(1) *Mechanical ball milling*: Matrix material and reinforcement particles are used as starting powders for the CS process. Powder is mixed according to weight or volume percentage and this mixture is placed in a long roll jar mill. Milling balls are added to this mixture as per the fixed powder-to-ball ratio. Jars are then rotated at certain speeds for a certain duration of time. After mixing, balls are to be removed from the feedstock through sieving and the feedstock is de-moisturized in the oven at a certain temperature and for a fixed duration of time. After drying, feedstock powder is vacuum sealed for CS operation. The feedstock preparation process is schematically illustrated in Figure 5.13.

(2) *Dry spraying*: The spray-drying method involves mixing fine powders in a slurry, followed by gas spraying/atomization and drying with hot gas, resulting in composite powders with a near-spherical shape and homogeneous structure. This approach is suitable for dispersing carbon nanotubes (CNTs) into metallic particles without causing structural damage to the CNTs. Therefore, it was employed to synthesize Al/CNT composite powders for CS deposition. However, the high-velocity impact during the deposition process caused severe plastic deformation and shearing effects, resulting in structural damage to the CNTs. Consequently, this method is not widely used to prepare composite powders for CS due to its low production efficiency and high cost [41].

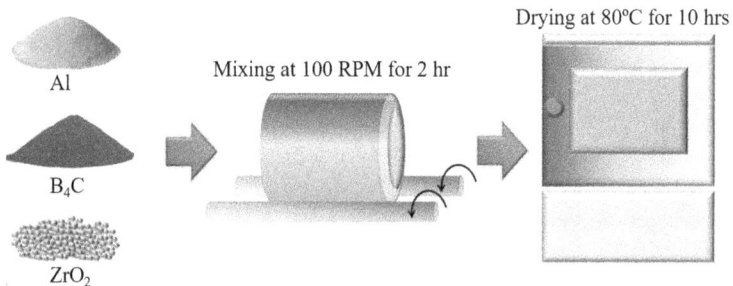

Drying at 80°C for 10 hrs

Al

Mixing at 100 RPM for 2 hr

B_4C

ZrO_2

Figure 5.13 Preparation of composite powder by the ball milling process [40].

(3) *Gas atomization*: Atomization is commonly considered the preferred method due to the desirable characteristics of the powder particles that can be achieved, making it ideal for AM. The process involves smelting metals or metal alloys in a protected crucible furnace until the composition of the metal liquids becomes homogeneous. The melted metal alloys are then refined and degassed before being poured into a gas nozzle, where a high-pressure gas stream disintegrates the liquid material into metal powder. Gas atomization, water atomization, and direct reduction are all reliant on industrial gases and gas technologies. Argon and nitrogen are required for gas atomization, whereas a hydrogen atmosphere is needed for water atomization, followed by oxide reduction [42].

5.2.2.3 Deposition geometry and reinforcement volume

Components or parts manufactured by the CSAM process have millimeter-level resolution. Manufacturing of curved and angled geometries using the CSAM process requires careful design of the part program because they change the standoff distance between the nozzle substrate and spray angle, which are critical parameters for the efficiency and quality of deposition [35]. The composites usually have lower reinforcement content compared to the initial feedstocks due to their relatively low deposition efficiency. Figure 5.14 displays four dotted lines that represent the fraction of the

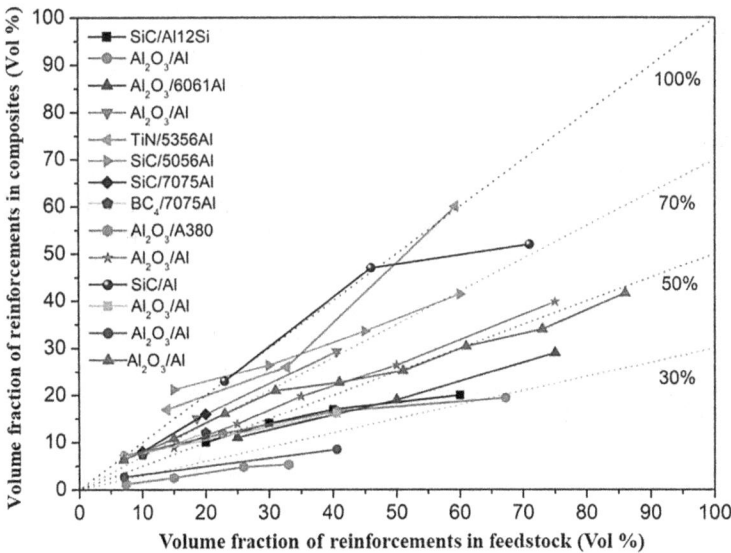

Figure 5.14 The volume fraction of reinforcement particles within the composites as a function of the initial powder mixtures [41].

reinforcement phase in the composite and the feedstocks. It is evident from these values that the deposition efficiency of the reinforcement particles lies in the range of 30% to 70% for most composite systems. However, some composite systems, such as Al/SiC, Al 5056/SiC, and Al 5356/TiN, exhibit very high deposition efficiency of reinforcements within the composite.

5.2.2.4 Microstructure and mechanical properties

Temperature generation during CSAM is less than the melting point of the feedstock material [7]. The CSAM process gives grain refinement due to dynamic recrystallization-induced severe plastic deformation of the feedstock material powder particles. But the extent of grain refinement within each deposition layer varies due to differences in the plastic deformation of feedstock material powder particles [43] i.e., grains which are closer to impacting the surface are finer. The CSAM-manufactured Al/Al_2O_3 MMC exhibits a dense structure with severely deformed Al particles and reinforcement particles embedded in it. The reinforcement particles are mainly located at the boundaries between the particles, and some of the larger ones are fractured into smaller fragments during deposition. The Al matrix adjacent to the Al_2O_3 particles is highly deformed, and the grains are significantly refined due to the enhanced peening effect [44]. The Al/TiB_2 composite has a heterogeneous structure with ultrafine grains formed at the inter-splat boundary regions where the particles were highly deformed, as shown in Figure 5.15. The large plastic deformation during the CSAM process resulted in dislocation networks and precipitation formation. Nano-sized TiB_2 particles are primarily dispersed along the grain boundaries of the Al matrix.

The incorporation of reinforcement particles in the feedstock results in a decrease in the porosity of composites due to the peening effects induced by the particles. Figure 5.16(a) shows that the porosity of the composite decreases as the reinforcement content increases and eventually reaches a constant low level, indicating a fully dense structure. The reduced porosity

Figure 5.15 SEM/EBSD orientation maps of the CSAM made (a) AlSi10Mg and (b) AlSi10Mg/TiB$_2$ composite [37].

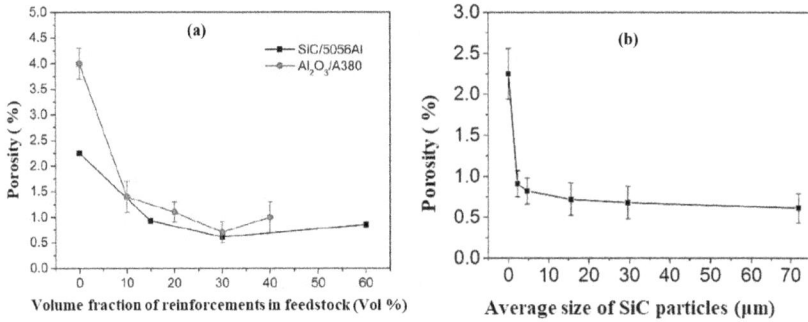

Figure 5.16 Porosity evolution as a function of (a) volume fraction of reinforcement particles in the feedstocks and (b) average SiC particle size [45, 47].

is attributed to the enhanced plastic deformation of the Al particles caused by the added reinforcement particles acting as peening particles. In addition to reinforcement content, the size and morphology of the reinforcement particles also influence the reduction of porosity. As demonstrated in Figure 5.16(b), the porosity of the Al 5056/SiC composite decreases slightly as the size of the reinforcement particle increases. Spherical reinforcement particles are more efficient in reducing porosity compared to irregularly shaped particles due to the enhanced in-situ peening effect of the spherical powders [45]. The CSAM manufactured parts have been found to contain fragments of broken oxides, resulting in 1–2% porosity with optimized process parameters. Post-AM heat treatment has been shown to improve the porosity [46].

Multi-layer depositions made by the CSAM have elongation less than 50% of the reference (i.e., with respect to wrought alloys). This is attributed to residual stresses and poor metallic bonding between the powder particles [7]. The presence of surface oxides on powder particles is one of the factors that reduce the strength and ductility of the CSAM-made depositions since they inhibit bonding among powder particles [48]. The CSAM-manufactured components from Ti-6Al-4V, and Inconel 718 have less strength and elongation than Al- and Cu-based alloys which exhibit higher relative ultimate tensile strength (UTS) and % elongation due to their higher ductility which helps in better bonding among the feedstock material particles. Tensile and compressive strength of as-sprayed depositions are relatively better than their properties after post-treatment which can be observed in Table 5.2.

5.2.3 Friction stir additive manufacturing process

The following subsections describe different aspects of the friction stir additive manufacturing (FSAM) process.

5.2.3.1 Working principle

The FSAM process uses the feedstock material in the form of sheets or laminates and bonds them together by friction stirring. The pin of a rotating tool is inserted between two consecutive sheets that are to be bonded together and traverses along the deposition direction as shown in Figure 5.17. The feedstock material which is in contact with the tool pin is heated and plasticized due to the friction between the tool and sheets. The generation of heat is due to the combined effect of friction and plastic deformation of sheets. A bond is created between the consecutive sheets due to the generated heat and movement of material from the leading edge of tool to the trailing edge [49].

Table 5.3 presents different MMCs manufactured by the FSAM process and subsequently major findings from these MMCs are briefly explained.

Various MMCs have been additively manufactured by the FSAM process. Venkit and Selvaraj [51] utilized the FSAM process to manufacture a six-layer single-track functionally gradient composite. They found that the microstructures along the build direction were non-uniform, and the stir zone was the most productive part of the build, having fine equiaxed grains. The varying thermal cycles along the build direction affected the grain sizes and precipitate sizes, resulting in an increase in strength and hardness of the FSAM build from the bottom layer to the top layer. Based on these results, the study suggests that the FSAM approach may be suitable for fabricating

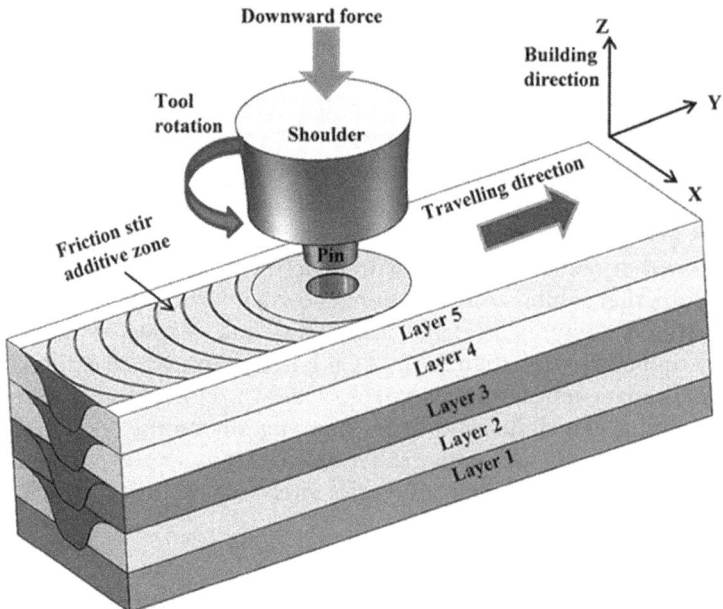

Figure 5.17 Schematic diagram depicting working principle of friction stir additive manufacturing process [50].

Table 5.3 Metal matrix composites manufactured by the friction stir additive manufacturing process

Feedstock material	Process parameters and number of deposition layers	Ultimate tensile strength (MPa)	Microhardness (HV)	Ref.
Alternative sheets of Al 6061-T6 and Al 7075-T6	Tool type: H13 tool having threaded, tapered, and conical pin Shoulder dia. of tool: 24 mm Root dia. of pin: 8 mm Tip dia. of pin: 6 mm Tool pin height: 4 mm Tool tilt angle: 2° Vertical load: 20 kN (constant) Tool rotational speed for Al 6061-T6 and Al 7075-T6: 1200 and 1100 rpm Tool traverse speed for Al 6061-T6 and Al 7075-T6: 40 and 50 mm/min No. of layers: 6	420	182.3	[51]
Alternative sheets of IF and St52 steel	Tool type: tungsten carbide cylindrical pin tool Shoulder dia. of tool: 20 mmTip dia. of pin: 6 mm Tool pin height: 0.5 mm Tool tilt angle: 3° Vertical load: 20 kN (constant) Tool rotational speed: 600 rpm Tool traverse speed: 40, 70 and 100 mm/min No. of layers: 2 (1st layer-St52 & 2nd layer-IF)	472	225	[52]
Sheets of AlZn alloy with Cu powder filled inside holes drilled in AlZn sheets	Tool type: cylindrical pin tool Shoulder dia. of tool: 18 mm Tip dia. of pin: 6 mm Tool pin height: 1.8 mm Tool tilt angle: 3° Tool rotational speed: 800 and 1200 rpm Tool traverse speed: 40, 70 and 100 mm/min No. of layers: 2	352.2	157	[53]

(Continued)

Table 5.3 (Continued)

Feedstock material	Process parameters and number of deposition layers	Ultimate tensile strength (MPa)	Microhardness (HV)	Ref.
Alternative sheets of 6061 alloy and A357-SiC	Tool tilt angle: 3° Tool rotational speed: 500–1000 rpm Tool traverse speed: 100–200 mm/min No. of layers: 2	-	170–180	[54]
Sheets of Al 6061 alloy with Al$_2$O$_3$ nano particles filled inside holes drilled in Al 6061 alloy sheets	Tool type: H13 tool steel Shoulder dia. of tool: 24 mm Root dia. of pin: 8 mm Tip dia. of pin: 6 mm Tool rotational speed: 1000 rpm Tool traverse speed: 100 mm/min No. of layers: 4	-	First layer: 71.9–73.1 Second layer: 79–88.5	[55]

large structures that are free of defects and have expected mechanical characteristics, making the newly fabricated composite a potential substitute for
conventional AA6061 material in automobile components to improve their
performance. Roodgari et al. [52] utilized the FSAM process to manufacture
a two-layer laminated composite steel. The results indicated that the layers
exhibited strong bonding, with a sharp interface observed at lower travel
speeds. However, as the traverse speed increased, diffusion occurred at the
interface, leading to the formation of a Widmanstatten ferrite and carbon
diffusion. This resulted in the highest hardness in the cross-section of the
composite, particularly in the stir zone and near the two steel interfaces. A
two-layer laminated composite was fabricated by Ardalanniya et al. [53],
using AlZn sheets as the matrix and Cu powder as the reinforcement. As
shown in Figure 5.18, holes were created between the two laminated AlZn

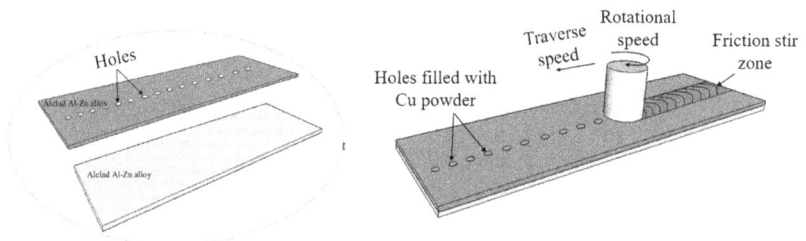

Figure 5.18 The schematic view of the FSAM method for the fabrication of the
laminated Al-Zn/Al-Zn-Cu$_{(p)}$ composite [53].

alloy sheets, and the reinforcement was filled in them. The results showed that the copper-rich reinforcement increased the ultimate tensile strength of the laminated composite by 8%, and the accumulation of copper-rich reinforcing particles near the interface of the laminated composite was the main cause of a sharp change in the hardness profile. Yan et al. [54] produced multifunctional metal matrix composites using the FSAM process, consisting of Al 6061 with more than 30% ceramic particulate additions of A357-SiC. The A357-SiC sheet was initially produced through casting and powder metallurgy. The inclusion of a high fraction of evenly distributed and extremely hard ceramic particles such as A357/SiC caused a significant reduction in interparticle spacing to a size comparable to the feedstock particle size, resulting in increased hardness comparable to some steels. Tan et al. [55] utilized the FSAM process to fabricate Al 6061/Al_2O_3 MMC. Figure 5.19 demonstrates the placement of the different sheets and their stirring during the process. The MMCs showed finer grains and higher hardness compared to Al 6061-T6 alloy. The re-stirring of the second layer resulted in a decreased aggregation of nanoparticles, leading to an increase in hardness and a decrease in grain size. The size of Al_2O_3 nanoparticles had a direct effect on the microstructure and hardness. As the nanoparticle size decreased from 100 to 10 nm, the grain size decreased from 14.8 to 12.1 μm, and the hardness increased from 67 to 87.3 HV.

5.2.3.2 Feedstock preparation methods

In the FSAM process, the matrix is in the form of sheets and reinforcements can be in either sheet or powder form. To incorporate reinforcement in the form of powders or flakes, channels or slots are first created in the substrate sheet. A second sheet, either of the same material as the substrate or a different material, is then placed over the channels or slots and processed by FSAM.

5.2.3.3 Deposition geometry

FSAM processes can achieve micron-level resolution. Complicated geometries having fine and unsupported features are generally beyond the capabilities of the FSAM processes due to the relatively high forces used to attain the deposition. Post-AM machining is often required to achieve these features.

5.2.3.4 Microstructure and mechanical properties

Temperature generation during FSAM generally lies between 0.6 and 0.9 T_m (where T_m is melting point of the feedstock material) [7]. Elongated grains with a certain rolling texture are typically present in the feedstock material (i.e., sheet) used in the FSAM process. Therefore, parts manufactured by them possess a non-homogenous microstructure consisting of elongated

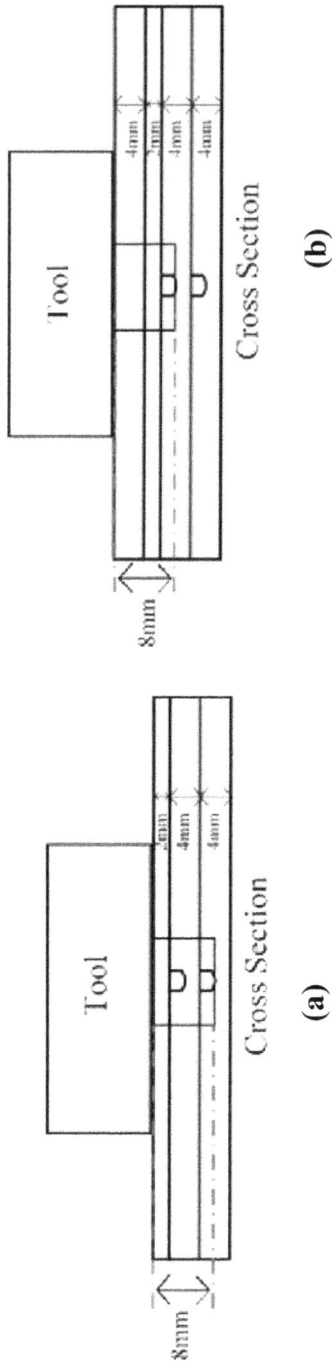

Figure 5.19 Schematic of FSAM process showing (a) stirring of 1st and 2nd layers and (b) re-stirring of 2nd layer [55].

grains in the bonded sheets and recrystallized fine grains at the stir zone. The deposition temperature is much higher in the FSAM than UAM processes which causes a large thermal gradient along the build direction and subsequently the formation of different phases. In addition, the presence of a tool pin in the FSAM process causes different microstructural zones from its center to a distance away from it, which also yields non-homogeneous microstructure (i.e., fine grains in stir zone and coarser in heat affected zone).

Microstructural kinetics during FSAM is impacted by various factors, such as the relative locations, different strains, and thermal cycles. Achieving desirable FSAM kinetics can result in highly favorable mechanical and fatigue properties, including improved formability and superplastic strengths, which are attributed to the development of equiaxed, recrystallized, and fine-grained microstructures. The material undergoes severe plastic deformation and recrystallization, leading to the formation of fine equiaxed grains within the stir zone of FSAM parts. Although material annealing at higher temperatures can negatively affect its mechanical characteristics, the grain refinement effect due to the development of refined grains is more prominent, resulting in enhanced microhardness of parts as per the Hall–Petch relationship. As a result, the tensile strength of the manufactured build also increases. Nevertheless, the presence of reinforcement volume percentage and size influences the microstructure and mechanical characteristics of composite materials. For 6061/A357-SiC MMCs, when the slot size (for filling reinforcements) in the Al 6061 matrix sheet was increased from 1.5 to 2 mm and 3 mm, the percentage volume of reinforcement also increased, as a result grain size decreased from 12.12 to 11.11 μm and 10.52 μm, respectively as shown in Figure 5.20. Similarly, reducing the reinforcement particle size from 100 nm to 50 nm, 25 nm, and 10 nm led to

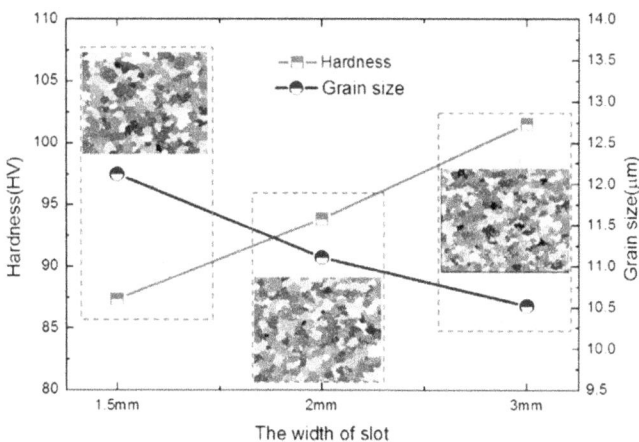

Figure 5.20 Effects of volume fraction on average grain size and hardness in the FSAM process [55].

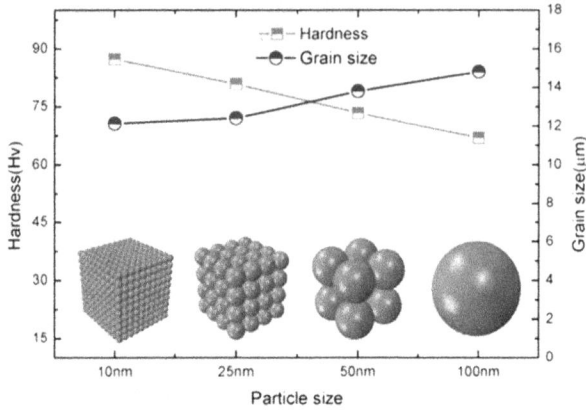

Figure 5.21 Effects of nanoparticle size on average grain size and hardness in the FSAM process [55].

a decrease in grain size from 14.81 to 13.79 μm, 12.4 μm, and 12.12 μm, as shown in Figure 5.21. This is because an increase in the volume fraction of reinforcement enhances the load transfer capacity from the matrix to the reinforcement. Furthermore, a higher volume fraction and the smaller size of reinforcement particles create obstacles for dislocation movement and increase the pinning effect, which restricts grain growth. Consequently, grain refinement is further promoted.

5.2.4 Additive friction stir deposition process

The following sub-sections describe different aspects of the additive friction stir deposition (AFSD) process.

5.2.4.1 Working principle

The AFSD process uses feedstock material in the form of rod or powder which is fed through a hollow non-consumable rotating tool as shown in Figure 5.22. Upon contact between the feedstock material and the substrate, the process of frictional heat generation initiates. This leads to a rise in temperature, causing the material to undergo plasticization. As the tool traverses along the deposition direction, the plasticized feedstock material extrudes to fill the space between the substrate and the rotating tool head. The bottom surface of the tool head performs the following two functions: controlling deposition layer thickness by providing a vertical constraint and ensuring good deposition quality by shearing the deposited material through its further rotation [18].

Table 5.4 presents different MMCs manufactured by the AFSD process and subsequently major findings from these MMCs are briefly explained.

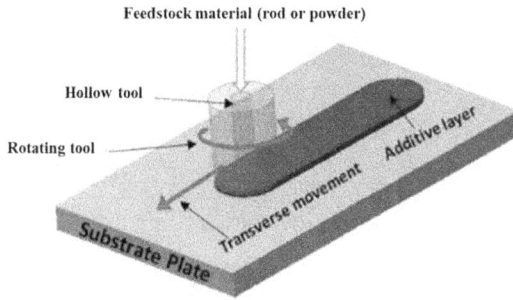

Figure 5.22 Schematic diagram depicting the working principle of the additive friction stir deposition process [19].

Table 5.4 Metal matrix composites manufactured by the additive friction stir deposition process

Feedstock material	Process parameters and type of deposition	Ultimate tensile strength (Mpa)	Microhardness (HV)	Ref.
A356/50%B$_4$C+50%MoS$_2$ rod	Type of tool: straight cylindrical flat tool Shoulder dia. Of tool: 20 mm Tool pin length: 3 mm Tool pin diameter: 6 mm Rotational speed: 1000 rpm Traverse speed: 50 mm/min	-	~450 for a MMC with 40 nm sized B4C particles	[56]
AA6351-T6/Al$_2$O$_3$ rod	Type of tools: straight cylindrical flat tools with (i) one hole and (ii) two holes Shoulder dia. Of tool: 19.05 mm Rotational speed: 3000 and 3500 rpm Traverse speed: 365, 400, 365 and 400 mm/min	-	By one hole: 70–75 By two holes: 75–80	[57]
Al 6063/8%ZrO$_2$ rod Al 6063/12% ZrO$_2$ rod	Type of tool: straight cylindrical flat tool Diameter of tool: 20 mmRotational speed: 1500 rpm Traverse speed: 60mm/min Rod plunging speed: 60 mm/min No. of layers: 3	281 313	77.6 92.2	[58]

(Continued)

Table 5.4 (Continued)

Feedstock material	Process parameters and type of deposition	Ultimate tensile strength (Mpa)	Microhardness (HV)	Ref.
Al 6061/20 vol.% SiC rod Al 6061/30 vol.% Mo rod Al 6061/6.1 vol.% W rod	Not reported	-	-	[59]
Al 6061-T6 square rod of size 1 cm filled with 2.5 wt.% graphene nanoplatelet (GNP) inside a bore having dia. Of 0.32 and depth of 7 cm.	Tool: 38.1 mm diameter cylindrical tool having squared bore with four teardrop protrusions Feedstock rod plunging speed: 72 mm/min Tool traverse speed: 72 mm/min Tool rotational speed: 300 rpm No. of layers: 1	-	45–70	[60]
AA6061-T6511 square rod of size 9.625 mm filled with 19.5 wt% SiC for simple tool (ST) and 20.3 wt% SiC for complex tool (CT) Cu-110 square rod of size 9.625 mm filled with 6.57 wt% SiC for ST and 5.76 wt% SiC for complex tool	Tool: 38.1 mm diameter cylindrical tool having squared bore (i) with, and (ii) without four teardrop protrusions Feedstock rod plunging speed: 25.2 mm/min Tool traverse speed: 76.2 mm/min Tool rotational speed: 300 rpm No. of layers: 1	-	-	[61]

Various MMCs have been additively manufactured by the AFSD process, including Al/MoS_2, Al/Al_2O_3, Al/ZrO_2, Al/SiC, Al/Mo, Al/W etc. Srinivasu et al. [56] manufactured a single-layer single track of Al A356/ B_4C+MoS_2 using the AFSD process and was then further treated using the FSP process. After FSP, microscopic examination indicated the breakage of coarse silicon needles and uniformly dispersed carbides in the A356 alloy matrix. The hardness of the surface composite layer was improved and homogeneous. Friction stir-treated composites with carbide particles demonstrated improved wear resistance. The use of solid lubricant MoS_2 powder considerably improved the wear resistance of the matrix material. Oliveira et al. [57] utilized the AFSD process to create a single-layer single-track deposition of Al $6061/Al_2O_3$ MMC for coating applications (see Figure 5.23). The AA6351/Al_2O_3 MMC demonstrated a lack of porosity and intermetallics in

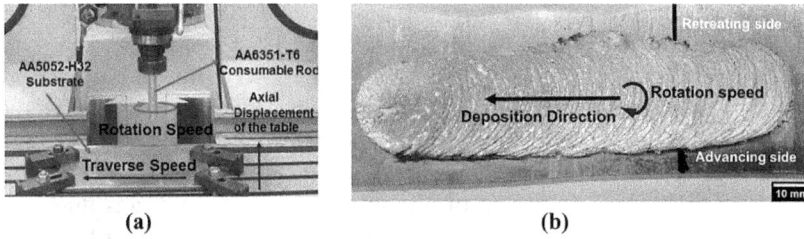

Figure 5.23 AFSD set up showing (a) consumable rod filled with ZrO_2 and (b) single-layer single-track deposition layer [57].

the retreating, center, and advancing regions. When the consumable feed-stock rod had one hole, the MMCs displayed a 7% increase in hardness compared to the deposition without particle addition. Meanwhile, the MMCs manufactured using two holes in the consumable feedstock rod exhibited a 7% higher hardness than those created using one hole, due to the higher volume fraction of reinforcement particles. However, there were still a considerable number of defects, which could be remedied by increasing the rod feed rates or axial loads. Both composites demonstrated equiaxed and refined grains. Patel et al. [58] used the AFSD process to produce a three-layer Al 6063/ZrO_2 composite through a single-track deposition, as illustrated in Figure 5.24. The study reported that the AFSD process resulted in good material flow, better mixing, and strong bonding between the layers. The microstructure exhibited fine grains, and the reinforced particles were uniformly distributed due to severe plastic deformation, dynamic recrystal-lization, and the pinning effect caused by the reinforced particles. The specific wear resistance was significantly improved compared to the as-received AA6063-T4 Al alloy rod. Based on their findings, the researchers concluded that the AFSD process holds great potential for manufacturing Al matrix composites with enhanced mechanical properties. Griffiths et al. [59] used

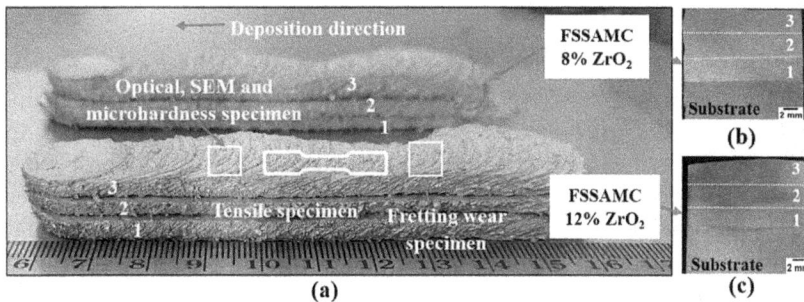

Figure 5.24 Additive friction stir deposition showing (a) three layers of Al 6063/ZrO_2 MMC, cross section of (b) MMC with 8% ZrO_2, and (c) MMC with 12% ZrO_2 [58].

the AFSD process to produce several types of MMCs, including Al 6061/20 vol.% SiC, Al 6061/30 vol.% Mo, and Al 6061/6.1 vol.% W. The study found that MMCs created using the AFSD process were free of porosity and possessed a uniform particle distribution. These favorable characteristics were attributed to the substantial material flow and mixing that occurred during the deposition process. The AFSD process is an attractive technology for the additive manufacturing of aluminum matrix composites due to its ease of use, variety of feed materials, scalability, and high quality of composites. Lopez et al. [60] utilized the AFSD process to create Al 6061/2.5 wt.% graphene nanoplatelet (GNP) MMCs. To manufacture the composites, AA6061 solid square rods in the T6 condition were cut into 1 cm × 1 cm × 9.5 cm blocks, which were then bored 7 cm in depth with a 0.32-cm-diameter hole for filling with GNP as reinforcement. The study found that the GNPs were distributed within and along grain boundaries, with their size reduced from 4,000 μm to 200 nm. The MMCs exhibited no porosity or delamination from the substrate. However, it was challenging to deposit a single layer or near-single layer of graphene necessary for strengthening using the AFSD process under the current processing conditions. Nonetheless, the process was successful in reducing carbide formation. Garcia et al. [61] used two different types of tools to manufacture Al 6061/x wt.% SiC and Cu/x wt.% SiC MMCs and investigated the effect of tool geometry on the distribution of reinforcement particles. The simple tool shape resulted in less uniform bulk dispersion of the reinforcement phase for Al-SiC samples, with localized areas of high SiC concentration (up to 69 area%). However, the Cu-SiC sample generated by the simple geometry tool exhibited a uniform distribution of reinforcement. On the other hand, complex geometry tools resulted in a more uniform distribution of SiC reinforcement throughout the deposit for both Al and Cu matrices, with localized reinforcement concentrations up to 42 and 28% area, respectively. All tool geometries produced samples with continuous contact between the reinforcement particle and matrix and excellent interfaces.

5.2.4.2 Feedstock preparation methods

The AFSD process offers a variety of options for feedstock materials, including rods, powder, and metal chips, as shown in Figure 5.25. Powders are prepared using the same method as mentioned in the CSAM process. To supply feedstock material in AFSD, one option is to use a solid AMC feed rod, which can be acquired or produced using traditional composite fabrication methods. Another option involves creating a feed rod with an aluminum alloy and filling it with loose reinforcement particles. During deposition, the packed particles blend into the metal matrix and the friction stir process enhances interface adhesion. Alternatively, matrix and reinforcement materials can be used in powder form as filler materials. The powders are provided via various hoppers, mixed inside the tool, and transported

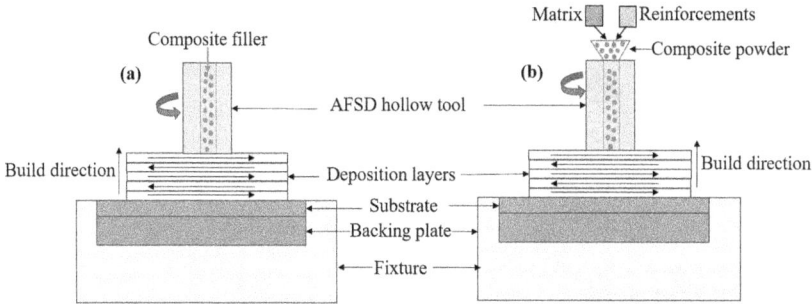

Figure 5.25 Demonstration of the feed stroke material approach (a) as rod and (b) as powder.

to the deposition zone. [59]. Moreover, Chaudhary et al. [62] reported that composite powders can be supplied straight to the deposition zone without being mixed in the tool. The feed material undergoes a thermal gradient due to the heat generated by dynamic contact friction, resulting in a high-temperature region above the deposition zone that aids in the consolidation of mixed powders before deposition.

5.2.4.3 Deposition geometry and reinforcement volume

Presently, AFSD processes have millimeter-level resolution. The parts manu-factured by the AFSD process often require post-AM machining to achieve complex features. Buckling under large tool forces is another factor in the AFSD that makes it difficult to manufacture tall and high aspect-ratio parts [63].

In order to ensure successful mixing, adhesion between interfaces, and continuous plastic deformation in the deposition zone, the softer matrix metal must flow around the harder reinforcement particles in AFSD when working with composites. However, increasing the amount of reinforcement can create obstacles to dislocation motion, due to factors such as Orowan stresses, strain gradients, and the accumulation of dislocations, which hin-der the required flow state for stirring [64]. Consequently, particles partially hinder material flow and increase the likelihood of agglomeration and void formation. Hence in the AFSD, the maximum volume fraction of the rein-forcement particles is limited to 30%. AA6351-T6/Al_2O_3 MMC showed heterogeneous distribution of reinforcing particles on both advancing and retreating sides, as shown in Figure 5.26. By contrast, in the case of Al 6063/ZrO2 MMC, reinforcement particles were uniformly distributed [58]. Moreover, for Al 6061/SiC, Al 6061/Mo and Al 6061/W MMCs, the rein-forcement particles were found to be distributed uniformly as reported by Griffiths et al. [59]. This implies that process parameters play a crucial role in deposition.

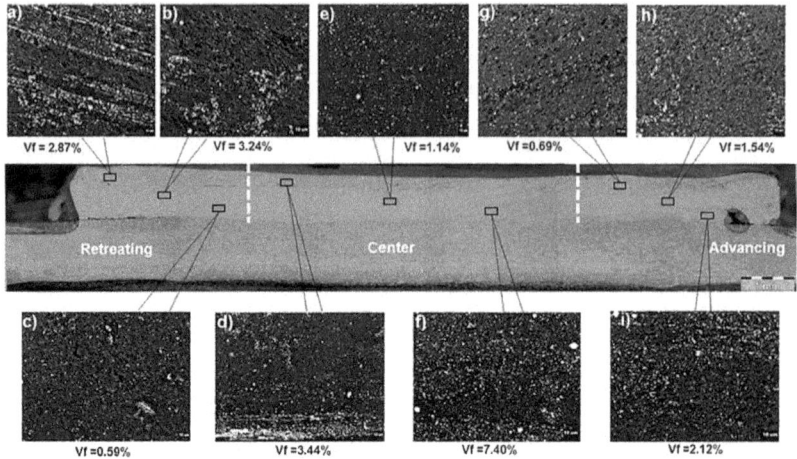

Figure 5.26 Macrographs of AA6351-T6/Al$_2$O$_3$ showing nonuniform distribution of Al$_2$O$_3$ particles in the (a–c) retreating side, (d–f) center and (g–i) advancing side [57].

5.2.4.4 Microstructure and mechanical properties

Temperature generation during AFSD generally lies between 0.6 and 0.9 T$_m$ (where, T_m is the melting point of the feedstock material) [7]. The use of feedstock material in rod or powder form in the AFSD process eliminates the use of a tool pin which results in fine, equiaxed, and homogeneous grains as the entire feedstock material experiences a similar temperature [20]. High temperatures generated during deposition in the AFSD process may lead to rapid grain growth in some materials which can be mitigated by controlling the rapid cooling of the deposition [65]. Parts manufactured by the AFSD process have shown much less porosity due to compressive forces. Some finely dispersed oxides have been observed in aluminum parts manufactured by the AFSD process [66, 67].

The majority of the composites manufactured by the AFSD process exhibit higher UTS and % elongation (above 50% of their wrought reference) in both as-deposited and heat-treated conditions. The grain growth at elevated temperatures during the AFSD process is absent in the CSAM and UAM processes, as their deposition temperature is lower [65]. Moreover, the microhardness of all composites was also found to be less than that observed in the matrix material. The AA5052-H32 plate substrate has a hardness value of about 72 HV0.2, while the AA6351-T6 consumable rod has a hardness value of 98.5 HV0.2. The hardness values between the substrate and rod in the as-received condition differ by roughly 36%. Furthermore, the hardness values was found to increase from the bottom layer towards the top layer in the majority of composites. It can be explained by the fact that the bottom layer undergoes heating effect several times as compared to the top layer which leads to a coarsening of the grains and reduces hardness [57].

5.3 CHALLENGES IN THE SOLID-STATE ADDITIVE MANUFACTURING OF COMPOSITES

There are several challenges associated with the use of solid-state additive manufacturing processes such as UAM, CSAM, FSAM, and AFSD in developing MMCs. Here are some of the major challenges identified with respect to each process:

5.3.1 Ultrasonic additive manufacturing (UAM)

- UAM is better suited for only thin foils and low-temperature working conditions
- It is challenging to achieve uniform distribution of the reinforcement particles in MMCs
- It is difficult to control the porosity and density of the MMCs
- It is challenging to ensure good bonding between the matrix material and reinforcement particles.
- It is important to establish appropriate process parameters for various materials and reinforcement types in order to scale up the process for large-scale manufacturing.
- The manufacturing of final components necessitates extensive machining, leading to an increase in production costs.

5.3.2 Cold spray additive manufacturing

- Variations in material characteristics remain an important challenge in CSAM. Inter-particle bonds and micro-pores, in particular, are prevalent deposit defects observed in CSAM-produced Cu, steel, and Ti composites. These defects may be caused by poor inter-particle bonding and insufficient plastic deformation during deposition, especially at low impact velocity.
- The combination of small grain structures and accumulating cold working effects can result in considerable residual stresses and embrittlement in the composite materials.
- It is challenging to get uniform distribution of reinforcement particles
- It is challenging to achieve high deposition rates for large-scale composite production
- It is difficult to control the thermal and mechanical stresses during deposition to avoid cracking or deformation

5.3.3 Friction stir additive manufacturing

- The presence of pin in FSAM tool greatly influences the temperature and reinforcement distribution which consequently leads to anisotropy in composites.
- The optimal process parameters greatly varies depending on the type of materials being used, the geometry of the reinforcement, and the desired properties of the composite.
- The current FSAM systems are limited in terms of the size of the components that can be produced, and the production rates are relatively low compared to other manufacturing processes. Therefore, there is a need to develop larger and more efficient FSAM systems to enable the production of larger components at higher production rates.
- The manufacturing of final components necessitates extensive machining, leading to an increase in production costs.

5.3.4 Additive friction stir deposition

- One of the critical challenges of AFSD is to achieve a uniform distribution of the reinforcement particles in the metal matrix. The parameters, such as the rotational and traverse speed of the tool, the volume fraction of reinforcement, and the feed rate of the reinforcement particles, can affect the distribution of the reinforcement particles.
- The AFSD process can result in a heterogeneous microstructure due to the inhomogeneous distribution of the reinforcement particles. It is essential to control the microstructure to achieve the desired properties.
- There is a need to optimize the process parameters to obtain high-quality MMCs. The optimization of parameters such as tool rotation speed, traverse speed, feed rate, and volume fraction of the reinforcement particles is critical for achieving the desired properties.
- The tool used in AFSD is subjected to high wear and tear due to the high temperature and mechanical stresses. Tool wear can lead to a decrease in the quality of the deposited material, and frequent tool maintenance can increase the cost of the process.
- The surface finish of the deposited material is crucial for the subsequent processing steps. The surface roughness can affect the adhesion of the next layer of the material, leading to delamination and poor quality.
- The selection of suitable matrix and reinforcement materials is critical for obtaining the desired properties of MMCs. The properties of the matrix and reinforcement materials, such as the coefficient of thermal expansion, melting point, and mechanical properties, should be compatible with each other to avoid thermal mismatch and deformation.

Overall, the challenges for each process are related to achieving a uniform distribution of the reinforcement particles, controlling the microstructure and properties of the deposited material, and optimizing the process parameters for different materials and reinforcement types. In addition, minimizing tool wear and maintenance requirements and scaling up the process for large-scale production are important considerations for all of these processes.

5.4 FUTURE DIRECTIONS

The field of additive manufacturing grew rapidly in the late twentieth century, and it is commonly expected that it will dominate practically every industrial area in the near future, accounting for around 80% of total output. Though SSAM processes have enormous potential in manufacturing MMCs, which prove difficult to produce using fusion-based additive manufacturing, much less work has been reported in this area. Therefore, the following are the essential requirements in the domain of composite manufacturing by SSAM processes.

- Scaling up the production of MMCs using UAM, CSAM, FSAM, and AFSD for industrial applications.
- Exploring the application of UAM, CSAM, FSAM, and AFSD for the production of MMCs with different types of reinforcement particles, such as carbon fibers, ceramic particles, and other nanomaterials
- There is a need for the further development and optimization of the process parameters to achieve high-quality and consistent MMCs with desirable properties. This includes exploring new materials and their combinations, as well as developing new equipment and processes that can improve the process efficiency and reduce production costs.
- In addition, there is a growing interest in the integration of these processes with digital manufacturing technologies such as AI and robotics, which can potentially enhance process control and automation. This can lead to the development of smart and functional MMCs with embedded sensors, actuators, and other electronic devices.
- Furthermore, the development of MMCs using these processes can also contribute to the sustainable and eco-friendly production of materials. The use of recycled and eco-friendly raw materials, as well as the reduction of waste and energy consumption, can be addressed by exploring and optimizing these processes for the production of MMCs.
- The UAM approach can be beneficial in embedding printed circuitries in a 3D way along the z-axis, perpendicular to the layup process, to enable the freeform integration of electrical circuitry into MMCs. Smart metal components with completely incorporated printed sensors, actuators, and micro- and nanoelectromechanical devices are among the potential applications.

REFERENCES

[1] Samal, P., Vundavilli, P.R., Meher, A., Mahapatra, M.M., (2020) Recent progress in aluminum metal matrix composites: A review on processing, mechanical and wear properties, *J Manuf Proc*. 59, 131–152. https://doi.org/10.1016/j.jmapro.2020.09.010

[2] Buragohain, M.K., *Composite Structures: Design, Mechanics, Analysis, Manufacturing, and Testing*, CRC Press, 2017.

[3] Patel, M., Murugesan, J., (2022) Effect of the tool pin eccentricity and cooling rate on microstructure, mechanical properties, fretting wear, and corrosion behavior of friction stir processed AA6063 alloy, *J Mater Eng Perform*. 31, 8554–8566. https://doi.org/10.1007/s11665-022-06860-y

[4] Sharma, S., *Composite Materials: Mechanics, Manufacturing and Modeling*, CRC Press, 2021.

[5] Rathee, S., Maheshwari, S., Siddiquee, A.N., Srivastava, M., (2017) A review of recent progress in solid-state fabrication of composites and functionally graded systems via friction stir processing, *Crit Rev Solid-State Mater Sci*. 43, 334–366. https://doi.org/10.1080/10408436.2017.1358146

[6] Kumar, S., Kar, A., (2021) A review of solid-state additive manufacturing processes, *Transact Indian Nat Acad Eng*. 6, 955–973. https://doi.org/10.1007/s41403-021-00270-7

[7] Tuncer, N., Bose, A., (2020) Solid-State metal additive manufacturing: A review, *JOM*. 72, 3090–3111. https://doi.org/10.1007/s11837-020-04260-y

[8] Galati, M., Iuliano, L., (2018) A literature review of powder-based electron beam melting focusing on numerical simulations, *Add Manuf*. 19, 1–20. https://doi.org/10.1016/j.addma.2017.11.001

[9] Zhuo, L., Wang, Z., Zhang, H., Yin, E., Wang, Y., Xu, T., Li, C., (2019) Effect of post-process heat treatment on microstructure and properties of selective laser melted AlSi10Mg alloy, *Mater Lett*. 234, 196–200. https://doi.org/10.1016/j.matlet.2018.09.109

[10] Lupone, F., Padovano, E., Ostrovskaya, O., Russo, A., Badini, C., (2021) Innovative approach to the development of conductive hybrid composites for selective laser sintering, *Compo Part A: App Sci Manuf*. 147, 106429. https://doi.org/10.1016/j.compositesa.2021.106429

[11] Zhang, C., Chen, F., Wang, Q., Liu, Y., Shen, Q., Zhang, L., (2023), Additive manufacturing and mechanical properties of TC4/Inconel 625 functionally graded materials by laser engineered net shaping, *Mater Sci Eng: A*. 862, 144370. https://doi.org/10.1016/j.msea.2022.144370

[12] Ishfaq, K., Abdullah, M., Mahmood, M. A., (2021), A state-of-the-art direct metal laser sintering of Ti6Al4V and AlSi10Mg alloys: Surface roughness, tensile strength, fatigue strength and microstructure, *Opt Laser Technol*. 143, 107366. https://doi.org/10.1016/j.optlastec.2021.107366

[13] Ren, L., Gu, H., Wang, W., Wang, S., Li, C., Wang, Z., Zhai, Y., Ma, P., (2020) The microstructure and properties of an Al-Mg-0.3Sc alloy deposited by wire arc additive manufacturing, *Metals*. 10, 320. https://doi.org/10.3390/met10030320

[14] Sawant, M.S., Jain, N.K., (2018) Investigations on additive manufacturing of Ti6Al4V by microplasma transferred arc powder deposition process, *J Manuf Sci Eng*. 140, 0811014. https://doi.org/10.1115/1.4040324

[15] Friel, R.J., Harris, R.A., (2013) Ultrasonic additive manufacturing – A hybrid production process for novel functional products, *Proc CIRP*. 6, 35–40. https://doi.org/10.1016/j.procir.2013.03.004

[16] Sunil, P., Gobinda, S., *Cold Spray in the Realm of Additive Manufacturing*, Springer, 2020.

[17] Phillips, B.J., Avery, D.Z., Liu, T., Rodriguez, O.L., Mason, C.J.T., Jordon, J.B., Brewer, L.N., Allison, P.G., (2019) Microstructure-deformation relationship of additive friction stir-deposition Al–Mg–Si, *Materialia*. 7, 100387. https://doi.org/10.1016/j.mtla.2019.100387

[18] Yu, H.Z., Jones, M.E., Brady, G.W., Griffiths, R.J., Garcia, D., Rauch, H.A., Cox, C.D., Hardwick, N., (2018) Non-beam-based metal additive manufacturing enabled by additive friction stir deposition, *Script Mater*. 153, 122–130. https://doi.org/10.1016/j.scriptamat.2018.03.025

[19] Kumar Srivastava, A., Kumar, N., Rai Dixit, A., (2021) Friction stir additive manufacturing – An innovative tool to enhance mechanical and microstructural properties, *Mater Sci Eng: B*. 263, 114832. https://doi.org/10.1016/j.mseb.2020.114832

[20] Mukhopadhyay, A., Saha, P., (2020) Mechanical and microstructural characterization of aluminium powder deposit made by friction stir based additive manufacturing, *J Mater Proc Technol*. 281, 116648. https://doi.org/10.1016/j.jmatprotec.2020.116648

[21] Wei, Y.K., Luo, X.T., Chu, X., Huang, G.S., Li, C.J., (2020) Solid-state additive manufacturing high performance aluminum alloy 6061 enabled by an in-situ micro-forging assisted cold spray, *Mater Sci Eng: A*. 776, 139024. https://doi.org/10.1016/j.msea.2020.139024

[22] Guo, H., Gingerich, M.B., Headings, L.M., Hahnlen, R., Dapino, M.J., (2019) Joining of carbon fiber and aluminum using ultrasonic additive manufacturing (UAM), *Comp Struct*. 208, 180–188. https://doi.org/10.1016/j.compstruct.2018.10.004

[23] Chen, X., Hehr, A., Dapino, M.J., Anderson, P.M., (2015) Deformation mechanisms in NiTi-Al composites fabricated by ultrasonic additive manufacturing, *Shape Mem Superelast*. 1, 294–309. https://doi.org/10.1007/s40830-015-0032-1

[24] Li, J., Monaghan, T., Nguyen, T.T., Kay, R.W., Friel, R.J., Harris, R.A., (2017) Multifunctional metal matrix composites with embedded printed electrical materials fabricated by ultrasonic additive manufacturing, *Compo Part B: Eng*. 113, 342–354. https://doi.org/10.1016/j.compositesb.2017.01.013

[25] Deng, Z., Gingerich, M.B., Han, T., Dapino, M.J., (2018) Yttria-stabilized zirconia-aluminum matrix composites via ultrasonic additive manufacturing, *Compo Part B: Eng*. 151, 215–221. https://doi.org/10.1016/j.compositesb.2018.06.001

[26] Hahnlen, R., Dapino, M.J., (2014) NiTi-Al interface strength in ultrasonic additive manufacturing composites, *Comp Part B: Eng*. 59, 101–108. https://doi.org/10.1016/j.compositesb.2013.10.024

[27] Li, D., (2021) A review of microstructure evolution during ultrasonic additive manufacturing, *Int J Adv Manuf Technol*. 113, 1–19. https://doi.org/10.1007/s00170-020-06439-8

[28] Kuo, C.H., Sridharan, N., Han, T., Dapino, M.J., Babu, S.S., (2019) Ultrasonic additive manufacturing of 4130 steel using Ni interlayers, *Sci Technol Weld Join*. 24, 382–390. https://doi.org/10.1080/13621718.2019.1607486

[29] Li, D., Soar, R., (2009) Influence of sonotrode texture on the performance of an ultrasonic consolidation machine and the interfacial bond strength, *J Mater Proc Technol.* 209, 1627–1634. https://doi.org/10.1016/j.jmatprotec.2008.04.018

[30] Sriraman, M.R., Gonser, M., Fujii, H.T., Babu, S.S., Bloss, M., (2011) Thermal transients during processing of materials by very high power ultrasonic additive manufacturing, *J Mater Proc Technol.* 211, 1650–1657. https://doi.org/10.1016/j.jmatprotec.2011.05.003

[31] Sridharan, N., Gussev, M.N., Parish, C.M., Isheim, D., Seidman, D.N., Terrani, K.A., Babu, S.S., (2018) Evaluation of microstructure stability at the interfaces of Al-6061 welds fabricated using ultrasonic additive manufacturing, *Mater Charact.* 139, 249–258. https://doi.org/10.1016/j.matchar.2018.02.043

[32] Shimizu, S., Fujii, H.T., Sato, Y.S., Kokawa, H., Sriraman, M.R., Babu, S.S., (2014) Mechanism of weld formation during very-high-power ultrasonic additive manufacturing of Al alloy 6061, *Acta Materialia.* 74, 234–243. https://doi.org/10.1016/j.actamat.2014.04.043

[33] Ward, A.A., Zhang, Y., Cordero, Z.C., (2018) Junction growth in ultrasonic spot welding and ultrasonic additive manufacturing, *Acta Materialia.* 158, 393–406. https://doi.org/10.1016/j.actamat.2018.07.058

[34] Sridharan, N., Norfolk, M., Babu, S.S., (2016) Characterization of Steel-Ta dissimilar metal builds made using very high power ultrasonic additive manufacturing (VHP-UAM), *Metal Mater Transact A.* 47, 2517–2528. https://doi.org/10.1007/s11661-016-3354-5

[35] Yin, S., Cavaliere, P., Aldwell, B., Jenkins, R., Liao, H., Li, W., Lupoi, R., (2018) Cold spray additive manufacturing and repair: Fundamentals and applications, *Add Manuf.* 21, 628–650. https://doi.org/10.1016/j.addma.2018.04.017

[36] Khomutov, M., Spasenko, A., Sova, A., Petrovskiy, P., Cheverikin, V., Travyanov, A., Smurov, I., (2021) Structure and properties of AA7075-SiC composite parts produced by cold spray additive manufacturing, *Int J Adv Manuf Technol.* 116, 847–861. https://doi.org/10.1007/s00170-021-07457-w

[37] Xie, X., Chen, C., Chen, Z., Wang, W., Yin, S., Ji, G., Liao, H., (2020) Achieving simultaneously improved tensile strength and ductility of a nano-TiB_2/AlSi10Mg composite produced by cold spray additive manufacturing, *Comp Part B: Eng.* 202, 108404. https://doi.org/10.1016/j.compositesb.2020.108404

[38] Klinkov, S.V., Kosarev, V.F., Shikalov, V.S., Vidyuk, T.M., (2022) optimization of cold spraying of neutron-absorbing (Al+B_4C) composite coatings, *J App Mech Tech Phy.* 63, 279–288. https://doi.org/10.1134/s0021894422020110

[39] Qiu, X., Tariq, N.U.H., Qi, L., Wang, J.Q., Xiong, T.Y., (2020) A hybrid approach to improve microstructure and mechanical properties of cold spray additively manufactured A380 aluminum composites, *Mater Sci Eng: A.* 772, 138828. https://doi.org/10.1016/j.msea.2019.138828

[40] Tariq, N.H., Gyansah, L., Wang, J.Q., Qiu, X., Feng, B., Siddique, M.T., Xiong, T.Y., (2018) Cold spray additive manufacturing: A viable strategy to fabricate thick B_4C/Al composite coatings for neutron shielding applications, *Surf Coat Technol.* 339, 224–236. https://doi.org/10.1016/j.surfcoat.2018.02.007

[41] Xie, X., Yin, S., Raoelison, R.N., Chen, C., Verdy, C., Li, W., Ji, G., Ren, Z., Liao, H., (2021) Al matrix composites fabricated by solid-state cold spray deposition: a critical review, *J Mater Sci Technol*, 86, 20–55. https://doi.org/10.1016/j.jmst.2021.01.026

[42] Unal, A., (2013) Production of rapidly solidified aluminium alloy powders by gas atomisation and their applications, *Powder Metallurgy.* 33, 53–64. https://doi.org/10.1179/pom.1990.33.1.53

[43] Lee, C., Kim, J., (2015) Microstructure of kinetic spray coatings: A review, *J Thermal Spray Technol*. 24, 592–610. https://doi.org/10.1007/s11666-015-0223-5

[44] Wang, Q., Birbilis, N., Huang, H., Zhang, M.X., (2013) Microstructure characterization and nanomechanics of cold-sprayed pure Al and Al-Al$_2$O$_3$ composite coatings, *Surf Coat Technol*. 232, 216–223. https://doi.org/10.1016/j.surfcoat.2013.05.009

[45] Wang, Y., Normand, B., Mary, N., Yu, M., Liao, H., (2017) Effects of ceramic particle size on microstructure and the corrosion behavior of cold sprayed SiCp/Al 5056 composite coatings, *Surf Coat Technol*. 315, 314–325. https://doi.org/10.1016/j.surfcoat.2017.02.047

[46] Wang, X., Feng, F., Klecka, M.A., Mordasky, M.D., Garofano, J.K., El-Wardany, T., Nardi, A., Champagne, V.K., (2015) Characterization and modeling of the bonding process in cold spray additive manufacturing, *Add Manuf*. 8, 149–162. https://doi.org/10.1016/j.addma.2015.03.006

[47] Qiu, X., Tariq, N.U.H., Wang, J.Q., Tang, J.R., Gyansah, L., Zhao, Z.P., Xiong, T.Y., (2018) Microstructure, microhardness and tribological behavior of Al$_2$O$_3$ reinforced A380 aluminum alloy composite coatings prepared by cold spray technique, *Surf Coat Technol*. 350, 391–400. https://doi.org/10.1016/j.surfcoat.2018.07.039

[48] Li, W.Y., Li, C.J., Liao, H., (2010) Significant influence of particle surface oxidation on deposition efficiency, interface microstructure and adhesive strength of cold-sprayed copper coatings, *App Surf Sci*. 256, 4953–4958. https://doi.org/10.1016/j.apsusc.2010.03.008

[49] Srivastava, M., Rathee, S., Maheshwari, S., Siddiquee, A.N., Kundra, T.K., (2019) A review on recent progress in solid-state friction based metal additive manufacturing: friction stir additive techniques, *Crit Rev Solid-State Mater Sci*. 44, 345–377. https://doi.org/10.1080/10408436.2018.1490250

[50] Palanivel, S., Nelaturu, P., Glass, B., Mishra, R.S., (2015) Friction stir additive manufacturing for high structural performance through microstructural control in an Mg based WE43 alloy, *Mater Des*. 65, 934–952. https://doi.org/10.1016/j.matdes.2014.09.082

[51] Venkit, H., Selvaraj, S.K., (2022) Novel technique for design and manufacture of alternating gradient composite structure of aluminum alloys using solid-state additive manufacturing technique, *Materials*. 15, 7369. https://doi.org/10.3390/ma15207369

[52] Roodgari, M.R., Jamaati, R., Jamshidi Aval, H., (2020) Fabrication of a 2-layer laminated steel composite by friction stir additive manufacturing, *J Manuf Proc*. 51, 110–121. https://doi.org/10.1016/j.jmapro.2020.01.031

[53] Ardalanniya, A., Nourouzi, S., Jamshidi Aval, H., (2021) Fabrication of the laminated Al-Zn-Cup/Al-Zn composite using friction stir additive manufacturing, *Mater Today Comm*. 27, 102268. https://doi.org/10.1016/j.mtcomm.2021.102268

[54] Yan, S., Chen, L., Yob, A., Renshaw, D., Yang, K., Givord, M., Liang, D., (2022) Multifunctional metal matrix composites by friction stir additive manufacturing, *J Mater Eng Perform*. 31, 6183–6195. https://doi.org/10.1007/s11665-022-07114-7

[55] Tan, Z., Li, J., Zhang, Z., (2021) Experimental and numerical studies on fabrication of nanoparticle reinforced aluminum matrix composites by friction stir additive manufacturing, *J Mater Res Technol*. 12, 1898–1912. https://doi.org/10.1016/j.jmrt.2021.04.004

[56] Srinivasu, R., Sambasiva Rao, A., Madhusudhan Reddy, G., Srinivasa Rao, K., (2015) Friction stir surfacing of cast A356 aluminium–silicon alloy with boron carbide and molybdenum disulphide powders, *Defence Technol.* 11, 140–146. https://doi.org/10.1016/j.dt.2014.09.004

[57] Oliveira, P.H.F., Galvis, J.C., Martins, J.D.P., Carvalho, A.L.M., (2017) Application of Friction surfacing to the production of aluminum coatings reinforced with Al2O3 particles, *Mater Res.* 20, 603–620. https://doi.org/10.1590/1980-5373-mr-2017-0039

[58] Patel, M., Chaudhary, B., Murugesan, J., Jain, N.K., (2022) Additive manufacturing of AA6063-ZrO_2 composite using friction stir surface additive manufacturing, *Transact Indian Insti Met,* 76, 581–588. https://doi.org/10.1007/s12666-022-02658-7

[59] Griffiths, R.J., Perry, M.E.J., Sietins, J.M., Zhu, Y., Hardwick, N., Cox, C.D., Rauch, H.A., Yu, H.Z., (2018) A perspective on solid-state additive manufacturing of aluminum matrix composites using MELD, *J Mater Eng Perform.* 28, 648–656. https://doi.org/10.1007/s11665-018-3649-3

[60] Lopez, J.J., Williams, M.B., Rushing, T.W., Confer, M.P., Ghosh, A., Griggs, C.S., Jordon, J.B., Thompson, G.B., Allison, P.G., (2022) A solid-state additive manufacturing method for aluminum-graphene nanoplatelet composites, *Materialia.* 23, 101440. https://doi.org/10.1016/j.mtla.2022.101440

[61] Garcia, D., Griffiths, R.J., Yu, H.Z., Additive friction stir deposition for fabrication of silicon carbide metal matrix composites, *Int Manuf Sci Eng Conf.,* 2020. https://doi.org/10.1115/MSEC2020-8532

[62] Chaudhary, B., Patel, M., Jain, N.K., Murugesan, J., Patel, V. (2023) Friction stir powder additive manufacturing of Al 6061/FeCoNi and Al 6061/Ni metal matrix composites: Reinforcement distribution, microstructure, residual stresses, and mechanical properties, *J Mater Proc Technol.* 319, 118061. https://doi.org/10.1016/j.jmatprotec.2023.118061

[63] Griffiths, R.J., Petersen, D.T., Garcia, D., Yu, H.Z., (2019) Additive friction stir-enabled solid-state additive manufacturing for the repair of 7075 aluminum alloy, *App Sci.* 9, 3486. https://doi.org/10.3390/app9173486

[64] Nan, C.W., Clarke, D.R., (1996) The influence of particle size and particle fracture on the elastic/plastic deformation of metal matrix composites, *Acta Materialia.* 44, 3801–3811.

[65] Garcia, D., Hartley, W.D., Rauch, H.A., Griffiths, R.J., Wang, R., Kong, Z.J., Zhu, Y., Yu, H.Z., (2020) In situ investigation into temperature evolution and heat generation during additive friction stir deposition: A comparative study of Cu and Al-Mg-Si, *Add Manuf.* 34, 101386. https://doi.org/10.1016/j.addma.2020.101386

[66] Rivera, O.G., Allison, P.G., Brewer, L.N., Rodriguez, O.L., Jordon, J.B., Liu, T., Whittington, W.R., Martens, R.L., McClelland, Z., Mason, C.J.T., Garcia, L., Su, J.Q., Hardwick, N., (2018) Influence of texture and grain refinement on the mechanical behavior of AA2219 fabricated by high shear solid-state material deposition, *Mater Sci Eng: A.* 724, 547–558. https://doi.org/10.1016/j.msea.2018.03.088

[67] Patel, M., Sangral, S., Murugesan, J., (2023) Study on microstructure and mechanical properties of hybrid aluminium matrix composite with ZrO_2 and Ni particle as reinforcement fabricated through FSP route. *Adv Compo Mater.* 32, 460–476. https://doi.org/10.1080/09243046.2022.2120729

Chapter 6

Role and capability of the Internet of Things in additive manufacturing and its application regimes

Sachin Kumar
NamTech Institute, Gandhinagar, India

Aditya Sharma
Dayalbagh Educational Institute, Agra, India

Jibin T. Philip
Indian Institute of Science, Bengaluru, India

6.1 INTRODUCTION

With the emergence of the fourth industrial revolution, popularly referred as Industry 4.0, there have been significant technological advancements, including the Internet of Things (IoT), the Industrial Internet of Things (IIoT), hyper-automation, and the distributed cloud. The IoT is about connecting the power of the internet with different devices beyond smartphones and computers in order to create seamless interfaces that will make our lives better. They work by sharing data with each other; this shared data can come from human behaviours and usage patterns. The key technologies that comprise the Industry 4.0 paradigm and have been previously implemented into the AM process possess immense potential to bring about miraculous manufacturing changes [1, 2]. The ongoing pandemic situation, too, has pushed the industries into the deployment of advanced technologies to deal effectively with the major challenges that have followed COVID-19 and also to keep up with customers' expectations. Therefore, manufacturing industries need to adopt the latest IoT into AM, especially with regard to lean and optimized manufacturing and the perceived proficiency of advanced printing technologies. This chapter identifies goals of sustainable manufacturing that can be accomplished as a result of the implementation of IoT in 3D printing technology [3]. Over the past few years, engineers have started to adopt 3D printing technologies in the manufacture of IoT devices. The benefit of using 3D printing in IoT is that there is flexibility in developing models which apply different scales, materials, specifications, and quantities. As IoT-enabling technology continues to mature and interest continues to grow, IoT systems are quickly being adopted in new

DOI: 10.1201/9781032616025-6

169

sectors and in new ways. The idea of human–machine interfaces (HMI) is also becoming increasingly significant. The digital modelling technique fosters intelligent humanistic machine integration by favourably influencing the designer's creative talents [4]. This internet-based technology is being widely tapped into by manufacturing processes [5]. The Internet of Behaviour (IoB) was created as a result of the Internet of Things (IoT), which significantly transformed how data is acquired and used for various industrial operations [6, 7]. IoT has been expanded in wide domains from manufacturing, welding, machining, biomedical, transportation, decision-making, and metal discovery [8–12]. Organizations that use these intelligent technologies and work with optimized solutions will advance in their development of resilience and flexibility, according to Gartner Strategic Technology Trends 2021 [13, 14].

On the other hand, organizations that do not quickly adopt these technologies risk falling behind. Successful pandemic management has been made possible by the widespread use of digital technology in 2020, especially in the medical sector [15–17]. The rising demand for various medical equipment has increased spending on IoT-enabled medical devices. Companies worldwide have now turned to the use of automated technologies to brace themselves for the potential uncertainties due to the lessons acquired from the COVID-19 pandemic. Manufacturing companies operating online are proven to be both time- and money-efficient. These businesses can manage risks using Industry 4.0's disruptive technology.

AM can facilitate creating complex designs more effectively with fewer material losses or modifications [10, 18–20]. Figure 6.1 illustrates how smart technologies such as artificial intelligence, drone technology, radio frequency identification, and predictive maintenance, have improved business processes in the manufacturing sector, including supply chain management, warehouse management systems, 3D printing, production and distribution systems, and racing and tracking [21, 22].

A delicate, lightweight item comprised of many materials may be created with the help of topology optimization and additive manufacturing (AM) [10, 23]. Large-scale end-to-end flow processing and the customization of manufactured goods are made possible by the use of IoT-enabled AM processes. Shape memory polymers, superabsorbent polyesters, and piezoelectric materials, among others, are stimuli-responsive manufacturing materials that may be produced in large quantities using this method.

Global industrialization and automation have pushed engineering systems to be precise, accurate, robust, and reliable. Moreover, the increasing demand for running accuracies necessitates the systems to be perfect. This can be achieved effectively if the systems are self-sufficient. The hybridization of 3D printing with IoT can effectively fulfill the purpose. The key motivations are:

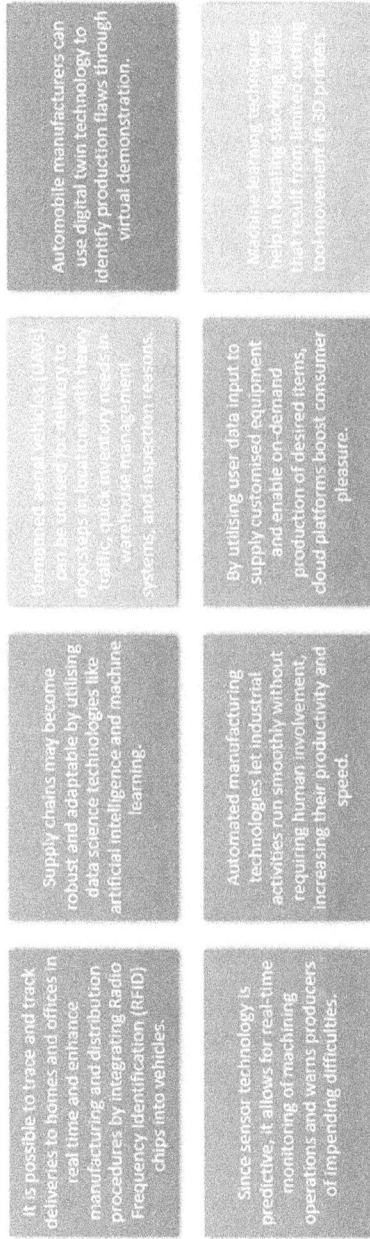

Application of Information technology in manufacturing industry

It is possible to trace and track deliveries to homes and offices in real time and enhance manufacturing and distribution procedures by integrating Radio Frequency Identification (RFID) chips into vehicles.

Supply chains may become robust and adaptable by utilising data science technologies like artificial intelligence and machine learning.

Unmanned aerial vehicles (UAVs) can be utilised for delivery to doorsteps in locations with heavy traffic, quick inventory needs, warehouse management systems, and inspection reasons.

Automobile manufacturers can use digital twin technology to identify production flaws through virtual demonstration.

Since sensor technology is predictive, it allows for real-time monitoring of machining operations and warns producers of impending difficulties.

Automated manufacturing technologies let industrial activities run smoothly without requiring human involvement, increasing their productivity and speed.

By utilising user data input to supply customised equipment and enable on-demand production of desired items, cloud platforms boost consumer pleasure.

Machine learning techniques help in locating stacking faults that result from limited cutting tool movement in 3D printers.

Figure 6.1 Emerging techniques in manufacturing industries [3].

1. 3D printing or AM, along with IoT, can produce typical and intricate shapes with high accuracy and precision, which are very challenging to produce by other methods.

2. 3D printing, in conjunction with IoT, can produce typical prototypes or working models, which can be demonstrated to the students or trainees and provide the hands-on experience.

3. In 3D printing, very little wastage is produced during and after the process; moreover, the generated waste can be effectively utilized for further production without compromising product quality. In addition, 3D printing and IoT hybridization may further reduce the generated waste. In other words, the process is highly efficient in terms of avoiding wastage.

4. 3D printing and IoT are relatively less time-consuming. Moreover, it can transfer artistic scene styles onto 3D reality with significantly less or no finishing process required for the fabricated products.

5. Very few human resources are required to regulate and operate the entire system efficiently.

For biomaterials with maximal strength and hardness such as polylactic acid, print speed is a significant machining parameter in 3D printing procedures [24–26]. Additionally, it demonstrates considerable skill to effectively utilize raw components and improve the quality of the finished output. One effective instrument that may be used for airport system repair procedures is internet-based additive manufacturing technology [27, 28]. Nevertheless, due to a lack of suitable infrastructure, the objective of automating AM has not been realized and put into practice. To fully grasp how IoT-enabled AM technology might change industrial operations, much more research has to be undertaken. To compare the benefits of IoT-enabled AM technologies and their potential for sustainable manufacturing, this chapter's primary goal is to assess the implementation potentials of internet-based technologies in the manufacturing sector. Furthermore, the critical features of modern AM technologies are briefly discussed through their essential application domains.

6.2 A SUSTAINABLE MANUFACTURING PLATFORM THROUGH INTERNET OF THINGS-ENABLED AM AND ITS POTENTIALS

Facilitated by the Internet of Things, AM assists with environmentally friendly production by reducing material usage and increasing the effectiveness of the process. Using materials that degrade over time, such as biodegradable ones, it is feasible to construct sustainable structures on a vast scale [3, 29, 30]. By automating these procedures, these application areas outperform traditional AM methods in terms of improving long-term economic viability. Additionally, automation technology gives AM a strong basis for

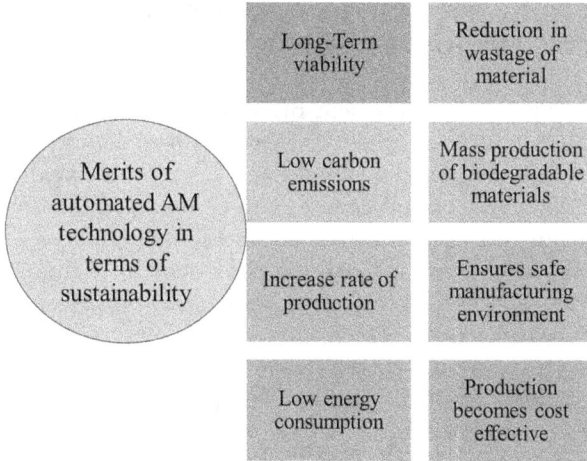

Figure 6.2 Sustainable automated AM manufacturing [3].

thriving in a cutthroat market climate by expanding their reach across different target audiences. It emphasizes lowering production costs and boosting output [31, 32].

Figure 6.2 shows how Internet-based technology and additive manufacturing might work together to maintain fast prototyping processes. AM technology may speed up time to market and manufacture tasks with intricate geometrical shapes. However, deploying a 3D printer in important businesses is difficult since fast prototyping procedures take a disproportionately long time to produce high-quality items. The AM procedure can potentially be improved to overcome this obstacle. Such contemporary changes contribute to tasks having a flawless surface finish, demonstrating the process's long-term effectiveness [33]. Robotic arms have been used by a number of businesses, including Concept Laser, Voodoo Manufacturing, 3D Systems, and Stratasys, to automate the rapid prototyping process. This led to lower operating expenses, a higher production rate, and higher sales of the final product. The printing machine can now run continuously without any need for human intervention, and productivity is now three times greater [33]. The efficiency of the printing process has also grown because of the IoT-enabled AM, as has the manufacturing life cycle [34].

Even though there are challenges in calculating the operating costs of AM processes, new research shows that the technologies are cost-effective for producing modest quantities. Making industrial operations more cost-effective requires the integration of IoT and additive manufacturing technologies. The use of AM processes for more diverse application areas has been facilitated by IoT-enabled AM technologies, which have significantly increased production volume and enabled mass manufacturing [2, 3].

6.3 INDUSTRIAL IMPLICATIONS OF IOT-ENABLED ADDITIVE MANUFACTURING

Manual data input is a component of conventional additive manufacturing operations. Additionally, due to their sophisticated understanding of 3D modeling design software, only knowledgeable users could exploit this technology profitably. The AM procedures will be operated by a novice, which leaves a possibility for mistakes and inaccurate data inputs. Future computer technology known as voice control has the potential to automate manual machine inputs, allowing even inexperienced users to execute 3D modeling and visualization using speech inputs [35]. Due to its stringent layer-by-layer integrated production process and uneven surface smoothness, 3D printing technology is often sluggish. The layer's size and fluctuation are the restrictions. Manufacturing flaws, including unintended interactions between tool extrusion and the project, constrained tool movement, and a lack of real-time monitoring, are to blame for the lack of a smooth job surface. An improved interface for the traditional AM techniques will make it more widely used and aid in the early detection of production flaws.

Additionally, this will support the identification of workplace flexibility, safe working conditions, and the conception of a 3D image of the final product [33, 36]. Those who rigorously adhere to a vegan diet sometimes miss out on essential nutrients in non-vegan food products. Vegans who don't enjoy eating foods high in protein like eggs, meat, and mutton can eat 3D-printed substitute foods that taste and feel like the real thing without risking their health or wellness. A vegan egg was developed by researchers using split green grammes and lentils.

The protein level of these vegan eggs was the same as that of the original food item. Automated 3D food printing technology may be used in restaurants to provide vegan options to their menus and quickly mass-produce customized vegan food products to satisfy customer demand [3, 37]. Because conventional house construction methods take longer than typical, advanced AM in the construction sector has become vital. Highly skilled laborers could be required for the building's construction. The safety of a building site is a serious concern since there are workers there [3].

Additionally, the real estate building contractor cannot monitor the construction site round the clock and ensure that the on-site workers are consuming the necessary quantities of cement and water. Automated AM processes have a significant role to play. It will speed up the building process and cut down on time. Additionally, less trash from the building will result from this. Furthermore, using humans during the construction process is not necessary. This promotes human safety and improves the dependability and reliability of the process. Utilizing on-demand manufacturing will also aid in the construction of the customized structure. As a result, utilizing an enhanced AM method, construction materials like concrete may also be 3D printed in large quantities [38–40].

Furthermore, automated 3D printing technologies will make it possible to produce fashion-related items that can react to environmental cues like humidity, light, and temperature in large quantities. These smart polymer-made clothing items shield the wearer from changes in the environment. The final client might provide feedback based on his preferred fashion clothing needs. A cloud interface may store the input data. Based on the data from the cloud-based platform, an appropriate AM technique may be selected for the effective manufacture of smart clothes. To satisfy client demands in the requested period and increase customer satisfaction, this AM process may be automated to mass commercialize desired items [41, 42].

To combat the current COVID-19 epidemic, personal protective equipment (PPE) is required. However, there are substantial issues with PPE use that we deal with daily, particularly for front-line healthcare personnel. PPE is pretty unpleasant in heat and high temperatures. Surgery mistakes might happen due to surgeons' more difficult hand movements caused by anxiety-related problems. Due to moisture that has collected and been deposited as vapors on goggles, it is challenging to maintain vocal conversations.

6.4 SIGNIFICANT APPLICATIONS AREAS OF INTERNET OF THINGS-ENABLED AM

Leading 3D printer producer Stratasys uses automated manufacturing techniques to mass-produce face shields utilizing a cloud-based platform to fend off COVID-19. More than 150 businesses and colleges participated in this programme [43]. When combined with automation, such novel AM methods have greater potential for mass manufacturing. By enabling the on-demand manufacture of materials, integrating AM with enterprise resource planning (ERP) platforms improves its capabilities [44]. The fields where the Internet of Things-enabled AM has been employed and those where it may someday be applied [3, 45–47] are included in Table 6.1.

The continuing COVID epidemic has severely disrupted the global supply chain, making it impossible for the world to meet its medical supply needs. Although there was a lack of personal protective equipment (PPE) and respirator machine components, AM service providers were able to address the immediate needs by delivering necessary medical services. It may now be imperative to move significantly from just-in-time operations to on-demand production. Eventually, supply chain changes would result from digital manufacturing technologies. By utilizing AM features, such as producing items with complex geometries, supply chain architectures may be optimized for production.

By minimizing material wastage, enhancing production flexibility, and facilitating decentralized manufacturing, automated AM can substantially influence supply chains. The automated AM helps to combine supply chain and product evolution, which helps in the evolution of the business model

Table 6.1 Application domains of IoT-enabled AM [3]

Applications	Description
Tissue engineering	Live tissue rings are delivered to an unmanned funnel guide for tissue engineering, where they are employed as raw materials to create bio-tubes during automated 3D printing. This provides immediate quality verification. This method may also be used to make hydrogels for tissue engineering [39].
Customization	AM is helpful in creating tailored items according to the needs of each particular consumer. This technique allows for the printing of complicated designs.
Defense	To analyze how AM settings affect the mechanical performance of the material, machine learning tools can be used. This technology helps AM procedures to be optimized [8, 9].
Shipyard industry	Steel and aluminum alloys are used in shipbuilding and are printed using additive manufacturing (AM) techniques. Older ship parts can be repaired using automated additive manufacturing (AM) technologies. Additionally, this technique may be utilized to print intelligent materials that can be used to create intelligent ship structural components. [41]
Automobile industry	Massive and medium-sized automotive parts are produced in large quantities using automated 3D printing technology. Minimizing human labor and enabling on-demand car component manufacture would improve supply chain and logistics operations across sectors. Machine components are built in accordance with consumer demands [42].
Medical industry	Human cells have been utilized to create miniature hearts utilizing AM. The use of bio-ink for 3D printing replicas of biological tissue is the subject of ongoing research into the application of AM in medicine. Large-scale automated 3D printing of customized, programmable human organs is possible.
Dentistry	The use of AM technology enables the production of numerous dental tools, components, and devices that are customized for each patient. Dentures with a crown and bridge may now be swiftly customized and produced.
Research and development	Today's AM relies heavily on applications, and researchers and developers play a big part in that. The technology platform itself is evolving and incorporating more recent developments. As a result, the teaching and learning process has room to grow. Students can use this and other emerging technologies to improve their knowledge.

as a whole. The automation of 3D printing technology gives supply networks a competitive advantage and improves the agility of supply chain procedures. The use of automation in additive manufacturing technologies is a major industry emphasis, mostly as the result of the importance placed on increasing customer value. The recent occurrence of the Suez Canal closure makes clear the complexity of global supply systems. Consequently, the secret to supply chain and logistics is lean manufacturing [48].

6.5 LIMITATIONS AND FUTURE SCOPE

Based on an extensive assessment of theoretical literature, the chapter identifies the potentiality of IoT-enabled AM processes through its crucial application areas. Although AM is a cutting-edge technology that is gaining traction in various sectors thanks to its many advantages, it is restricted by a number of problems that slow down its rapid development. The use of this technology has not yet been realized, however. The research material is insufficient because this field of study is new. After the pandemic, information technology will still be used in industrial companies.

Cost is essential because AM materials and equipment are prohibitively expensive, which is a significant barrier to using AM technology in mass production. Metrological standards for AM components should have been improved [49, 50]. Information technology and computer-controlled, computer-driven equipment are used to execute Industry 4.0. A single control unit may be used to adjust the processes of various AM machines that are computer-controlled equipment. This technology makes it possible to integrate and manage several pieces of equipment in a facility from a remote setting. This combination enables each machine to produce a particular product to the user [51–53]. Due to their versatility in creating many commodities concurrently with virtually infinite levels of complexity, AM machines are an inevitable part of the modern industrial era. The development of AM methods demonstrates their potential to revolutionize business structures and industrial industries. The development of global supply chain operations and the provision of high-quality services would both be facilitated by this technology. The subject of IoT in additive manufacturing is selected while considering the potential advantages of IoT-enabled fast prototyping techniques and bearing in mind the industries' experiences throughout the pandemic.

6.6 CONCLUSIONS

Over the past 200 years, there have been several industrial revolutions that have changed the manufacturing environment. Despite this, the so-called "Industry 4.0" revolution has shown a revolutionary integrated production system built on cutting-edge in-forming technology. Intelligent manufacturing makes it possible to use a highly adaptable manufacturing method to switch between customized, high-quality products and personalized mass production swiftly.

The significance of integrating IoT with additive manufacturing processes is the main topic of this chapter. Significant application areas have been determined through a literature-based examination of automated AM processes, their characteristics, and their viability. With the help of this technology, intelligent materials that react instantly to environmental changes may be printed. Large-scale intelligent designs may be created using these flexible materials. Tissue rings can also be used as raw materials to create human organ replicas. Due to its inability to mass-produce final items in a shorter amount of time, AM cannot completely replace traditional manufacturing techniques. IoT, on the other hand, expands the vast technical capabilities of fast prototyping methods, making it possible to manufacture customized goods on demand.

The implementation of intelligent technologies would be beneficial for AM procedures that need to produce complicated components. With less human participation and fewer mistakes, IT may increasingly substitute manual processes. The job's surface quality is enhanced, and manufacturing flaws are eradicated with artificial intelligence and machine learning. Implementing IoT in AM will promote lean, optimized, and sustainable production. By embracing lean manufacturing, decreasing wasteful material input, streamlining production procedures, increasing production flexibility, and enabling on-demand manufacturing, automated AM contributes to improving supply chain operations. IoT-enabled AM processes are helping decentralized supply chains. An increase in the use of information technologies such as the Internet of Things, artificial intelligence, and machine learning for automating fast prototyping procedures for mass-producing medical equipment has also been attributed to the COVID-19 pandemic. Lean manufacturing is one of the answers, as evidenced by the gaps in our supply chain systems revealed by the spreading epidemic.

The following points can be taken into account to establish future research directions for IoT and AM:

1. Artificial limbs: Human and animals can lose their limbs in a variety of different ways. Although a variety of off-the-shelf limbs are currently available on the market, these limbs are rarely a particularly comfortable fit. 3D printing and IoT may be the best methodology to produce the artificial limb with the required attributes in less time [54]. Moreover, as mentioned earlier, it will have less negligible waste if the rework is needed.

2. Optometry: Many of the human population use spectacles, goggles, or sunglasses. These spectacles are available in standard sizes and shapes. Similar to the case with limbs, not all humans have the same facial structure. With the advancement of 3D printing, along with the IoT, it will be possible to produce customized spectacles which will be more comfortable for the wearers.

3. Portable houses: In various circumstances, such as natural disasters and military operations and construction sites (roads, buildings, dams, etc.), portable houses are required, in order to provide shelter. In these

situations, portable houses are made with conventional methods. These houses are dumped/destroyed after their use comes to an end and new ones are then reconstructed at another site. This has costs in terms of not only money and the labour involved but also in terms of the natural resources involved. In these situations, portable and movable houses may be built to avoid and reduce wastage. The cost of these houses may seem to be unjustified in the short term, but they may prove a good long-term investment. Recently, the Indian Army has built and deployed such types of houses [55].

4. Coated pots for kids: Metallic, glass, or ceramic-based pots and utensils are used in various household applications. These pots usually fall from the grips of toddlers, kids, children, and aged persons and get dented. After that, these pots lost their appearance. If these pots are coated with non-metallic materials using 3D printing and IoT, meaning not only that they are prevented from dents, but also that the grip will be improved.

5. Decorative household applications: Various delicate and fragile decorative items such as electrical switches, lampshades, showpieces, etc., which are usually made of metals, glass, or ceramics, may be manufactured by non-metallic materials using 3D printing and IoT. It will increase the life of such items and reduce wastage.

REFERENCES

[1] Choudhary S, Islam A, Mukherjee B, Richter J, Arold T, Niendorf T, et al. Plasma sprayed Lanthanum zirconate coating over additively manufactured carbon nanotube reinforced Ni-based Composite: Unique performance of thermal barrier coating system without bondcoat. *Appl Surf Sci* 2021;550:149397. https://doi.org/10.1016/J.APSUSC.2021.149397

[2] Haghnegahdar L, Joshi SS, Dahotre NB. From IoT-based cloud manufacturing approach to intelligent additive manufacturing: industrial Internet of Things—an overview. *Int J Adv Manuf Technol* 2022;119:1461–78. https://doi.org/10.1007/s00170-021-08436-x

[3] Ashima R, Haleem A, Javaid M, Rab S. Understanding the role and capabilities of Internet of Things-enabled Additive Manufacturing through its application areas. *Adv Ind Eng Polym Res* 2022;5:137–42. https://doi.org/10.1016/j.aiepr.2021.12.001

[4] Zboinska MA. Influence of a hybrid digital toolset on the creative behaviors of designers in early-stage design. *J Comput Des Eng* 2019;6:675–92. https://doi.org/10.1016/j.jcde.2018.12.002

[5] Nascimento DLM, Alencastro V, Quelhas OLG, Caiado RGG, Garza-Reyes JA, Lona LR, et al. Exploring Industry 4.0 technologies to enable circular economy practices in a manufacturing context: A business model proposal. *J Manuf Technol Manag* 2018;30:607–27. https://doi.org/10.1108/JMTM-03-2018-0071

[6] Zhong RY, Dai QY, Qu T, Hu GJ, Huang GQ. RFID-enabled real-time manufacturing execution system for mass-customization production. *Robot Comput Integr Manuf* 2013;29:283–92. https://doi.org/10.1016/j.rcim.2012.08.001

[7] Hazen BT, Boone CA, Ezell JD, Jones-Farmer LA. Data quality for data science, predictive analytics, and big data in supply chain management: An introduction to the problem and suggestions for research and applications. *Int J Prod Econ* 2014;154:72–80. https://doi.org/10.1016/j.ijpe.2014.04.018

[8] Kumar S, Gaur V, Wu CS. Machine learning for intelligent welding and manufacturing systems: research progress and perspective review. *Int J Adv Manuf Technol* 2022 2022:1–29. https://doi.org/10.1007/S00170-022-10403-Z

[9] Kumar S, Gopi T, Harikeerthana N, Gupta MK, Gaur V, Krolczyk GM, et al. Machine learning techniques in additive manufacturing: a state of the art review on design, processes and production control. *J Intell Manuf* 2022;1–35. https://doi.org/10.1007/s10845-022-02029-5

[10] Kumar S, Kar A. A review of solid-state additive manufacturing processes. *Trans Indian Natl Acad Eng* 2021;6:955–73. https://doi.org/10.1007/S41403-021-00270-7

[11] Kumar S, Wu C. Eliminating intermetallic compounds via Ni interlayer during friction stir welding of dissimilar Mg/Al alloys. *J Mater Res Technol* 2021;15:4353–69. https://doi.org/10.1016/J.JMRT.2021.10.065

[12] Kumar S, Wu C. Strengthening effects of tool-mounted ultrasonic vibrations during friction stir lap welding of Al and Mg alloys. *Metall Mater Trans A Phys Metall Mater Sci* 2021;52:2909–25. https://doi.org/10.1007/s11661-021-06282-w

[13] Li M, Jin Z, C Chen. Application of RFID on products tracking and tracing system. Comput Integr *Manuf Syst* 2010;16:202–8.

[14] Raman S, Patwa N, Niranjan I, Ranjan U, Moorthy K, Mehta A. Impact of big data on supply chain management. *Int J Logist Res Appl* 2018;21:579–96. https://doi.org/10.1080/13675567.2018.1459523

[15] Xu L, Kamat VR, Menassa CC. Automatic extraction of 1D barcodes from video scans for drone-assisted inventory management in warehousing applications. *Int J Logist Res Appl* 2018;21:243–58. https://doi.org/10.1080/136755 67.2017.1393505

[16] Kritzinger W, Karner M, Traar G, Henjes J, Sihn W. Digital Twin in manufacturing: A categorical literature review and classification. *IFAC-PapersOnLine*, vol. 51, 2018, pp. 1016–22. https://doi.org/10.1016/j.ifacol.2018.08.474

[17] Lee J, Lapira E, Bagheri B, Kao H. Recent advances and trends in predictive manufacturing systems in big data environment. *Manuf Lett* 2013;1:38–41. https://doi.org/10.1016/j.mfglet.2013.09.005

[18] Choudhary S, Pandey A, Gaur V. Role of microstructural phases in enhanced mechanical properties of additively manufactured IN718 alloy. *Mater Sci Eng A* 2022:144484. https://doi.org/10.1016/J.MSEA.2022.144484

[19] Kar A, Kumar S, Kailas SV. Developing multi-layered 3D printed homogenized structure using solid state deposition method. *Mater Charact* 2023;199:112770. https://doi.org/10.1016/J.MATCHAR.2023.112770

[20] Kumar S, Gaur V. *Advances in Fatigue Prediction Techniques*. Adv. Fatigue Fract. Test. Model., Intech Open London; 2022, pp. 01–15.

[21] Lee J, Lapira E, Yang S, Kao A. Predictive manufacturing system - Trends of next-generation production systems. IFAC Proc. vol. 46, IFAC Secretariat; 2013, pp. 150–6. https://doi.org/10.3182/20130522-3-BR-4036.00107

[22] Parmar H, Khan T, Tucci F, Umer R, Carlone P. Advanced robotics and additive manufacturing of composites: Towards a new era in Industry 4.0. *Mater Manuf Process* 2022;37:483–517. https://doi.org/10.1080/10426914.2020.1866195

[23] Li W, Cao C, Yin S. Solid-state cold spraying of Ti and its alloys: A literature review. *Prog Mater Sci* 2020;110:100633. https://doi.org/10.1016/j.pmatsci.2019.100633

[24] Durga Prasad Reddy J, Mishra D, Chetty N. Strength and Hardness of 3D Printed Poly Lactic Acid and Carbon Fiber Poly Lactic Acid Thermoplastics. *Springer Proc Mater* 2020;8:625–34. https://doi.org/10.1007/978-981-15-7827-4_64

[25] Kumar S, Wu CS. Review: Mg and Its Alloy - Scope, Future Perspectives and Recent Advancements in Welding and Processing. *J Harbin Inst Technol* 2017;24:1–37. https://doi.org/10.11916/j.issn.1005-9113.17065

[26] Kumar S, Wu CS, Padhy GK, Ding W. Application of ultrasonic vibrations in welding and metal processing: A status review. *J Manuf Process* 2017;26:295–322. https://doi.org/10.1016/j.jmapro.2017.02.027

[27] Ceruti A, Marzocca P, Liverani A, Bil C. Maintenance in aeronautics in an Industry 4.0 context: The role of augmented reality and additive manufacturing. *J Comput Des Eng* 2019;6:516–26. https://doi.org/10.1016/j.jcde.2019.02.001

[28] Kumar S, Kishor B. Ultrasound Added Additive Manufacturing for Metals and Composites: Process and Control. Addit. Subtractive Manuf. Compos., Springer, Singapore; 2021, pp. 53–72. https://doi.org/10.1007/978-981-16-3184-9_3

[29] Kumar S, Wu CS, Shi L. Intermetallic diminution during friction stir welding of dissimilar Al/Mg alloys in lap configuration via ultrasonic assistance. *Metall Mater Trans A* 2020;51:5725–42. https://doi.org/10.1007/s11661-020-05982-z

[30] Kumar S, Wu CS. Suppression of intermetallic reaction layer by ultrasonic assistance during friction stir welding of Al and Mg based alloys. *J Alloys Compd* 2020;827:154343. https://doi.org/10.1016/j.jallcom.2020.154343

[31] Machado CG, Winroth MP, Ribeiro da Silva EHD. Sustainable manufacturing in Industry 4.0: an emerging research agenda. *Int J Prod Res* 2020;58:1462–84. https://doi.org/10.1080/00207543.2019.1652777

[32] Althammer F, Ruf F, Middendorf P. Size optimization methods to approximate equivalent mechanical behaviour in thermoplastics. *J Comput Des Eng* 2021;8:170–88. https://doi.org/10.1093/jcde/qwaa069

[33] Urhal P, Weightman A, Diver C, Bartolo P. Robot assisted additive manufacturing: A review. *Robot Comput Integr Manuf* 2019;59:335–45. https://doi.org/10.1016/j.rcim.2019.05.005

[34] Chen B, Wan J, Shu L, Li P, Mukherjee M, Yin B. Smart factory of Industry 4.0: Key technologies, application case, and challenges. *IEEE Access* 2017;6:6505–19. https://doi.org/10.1109/ACCESS.2017.2783682

[35] Allen P, Feiner S, Troccoli A, Benko H, Ishak E, Smith B. Seeing into the past: Creating a 3D modeling pipeline for archaeological visualization. *Proc 2nd Int Symp 3D Data Process Vis Transm* 2004;751–8. https://doi.org/10.1109/TDPVT.2004.1335391

[36] Bhatt PM, Peralta M, Bruck HA, Gupta SK. Robot assisted additive manufacturing of thin multifunctional structures. *ASME 2018 13th Int. Manuf. Sci. Eng. Conf. MSEC 2018*, vol. 1, American Society of Mechanical Engineers (ASME); 2018. https://doi.org/10.1115/MSEC2018-6620

[37] Anukiruthika T, Moses JA, Anandharamakrishnan C. 3D printing of egg yolk and white with rice flour blends. *J Food Eng* 2020;265:109691. https://doi.org/10.1016/j.jfoodeng.2019.109691

[38] Buchanan C, Gardner L. Metal 3D printing in construction: A review of methods, research, applications, opportunities and challenges. *Eng Struct* 2019;180:332–48. https://doi.org/10.1016/j.engstruct.2018.11.045

[39] Tay YWD, Panda B, Paul SC, Noor Mohamed NA, Tan MJ, Leong KF. 3D printing trends in building and construction industry: A review. *Virtual Phys Prototyp* 2017;12:261–76. https://doi.org/10.1080/17452759.2017.1326724

[40] Hager I, Golonka A, Putanowicz R. 3D printing of buildings and building components as the future of sustainable construction? *Procedia Eng* 2016;151:292–9. https://doi.org/10.1016/j.proeng.2016.07.357

[41] Yap YL, Yeong WY. Additive manufacture of fashion and jewellery products: A mini review: This paper provides an insight into the future of 3D printing industries for fashion and jewellery products. *Virtual Phys Prototyp* 2014;9:195–201. https://doi.org/10.1080/17452759.2014.938993

[42] Hashemi Sanatgar R, Cayla A, Campagne C, Nierstrasz V. Manufacturing of polylactic acid nanocomposite 3D printer filaments for smart textile applications. *IOP Conf. Ser. Mater. Sci. Eng.*, vol. 254, Institute of Physics Publishing; 2017, p. 072011. https://doi.org/10.1088/1757-899X/254/7/072011

[43] Mumbai-based startup 3D prints protective face shields for doctors | Technology News, The Indian Express n.d. https://indianexpress.com/article/technology/tech-news-technology/covid-19-mumbai-firm-3d-prints-face-shields-for-doctors-in-city-6337402/ (accessed August 30, 2022).

[44] 3D Printing Community responds to COVID-19 and Coronavirus resources - 3D Printing Industry n.d. https://3dprintingindustry.com/news/3d-printing-community-responds-to-covid-19-and-coronavirus-resources-169143/ (accessed August 30, 2022).

[45] Alavi AH, Jiao P, Buttlar WG, Lajnef N. Internet of Things-enabled smart cities: State-of-the-art and future trends. *Meas J Int Meas Confed* 2018;129:589–606. https://doi.org/10.1016/j.measurement.2018.07.067

[46] Geetha S, Gouthami S. Internet of things enabled real time water quality monitoring system. *Smart Water* 2016;2:1. https://doi.org/10.1186/s40713-017-0005-y

[47] Guimarães DA, Pereira EJT, Alberti AM, Moreira JV. Design guidelines for database-driven internet of things-enabled dynamic spectrum access. *Sensors* 2021;21:3194. https://doi.org/10.3390/s21093194

[48] Verboeket V, Krikke H. Additive manufacturing: A game changer in supply chain design. *Logistics* 2019;3:13. https://doi.org/10.3390/logistics3020013

[49] Yu C, Jiang J. A perspective on using machine learning in 3D bioprinting. *Int J Bioprinting* 2020;6:4–11. https://doi.org/10.18063/ijb.v6i1.253

[50] Verboeket V, Khajavi SH, Krikke H, Salmi M, Holmstrom J. Additive manufacturing for localized medical parts production: A case study. *IEEE Access* 2021;9:25818–34. https://doi.org/10.1109/ACCESS.2021.3056058

[51] Kleer R, Piller FT. Local manufacturing and structural shifts in competition: Market dynamics of additive manufacturing. *Int J Prod Econ* 2019;216:23–34. https://doi.org/10.1016/j.ijpe.2019.04.019

[52] Li Y, Jia G, Cheng Y, Hu Y. Additive manufacturing technology in spare parts supply chain: a comparative study. *Int J Prod Res* 2017;55:1498–515. https://doi.org/10.1080/00207543.2016.1231433

[53] Heinen JJ, Hoberg K. Assessing the potential of additive manufacturing for the provision of spare parts. *J Oper Manag* 2019;65:810–26. https://doi.org/10.1002/joom.1054

[54] Ahuett-Garza H, Kurfess T. A brief discussion on the trends of habilitating technologies for Industry 4.0 and Smart manufacturing. *Manuf Lett* 2018;15:60–3. https://doi.org/10.1016/J.MFGLET.2018.02.011

[55] 3D printed houses: Indian Army's first ever digitally created accommodation. Read here | Mint n.d. https://www.livemint.com/news/india/3d-printed-houses-indian-army-s-first-ever-digitally-created-accommodation-read-here-11647264160897.html (accessed November 26, 2022).

Chapter 7

Assessment of defects in solid-state additive manufacturing through conventional methods and sensor-based data

Gaurav Kishor, Raju Prasad Mahto and Krishna Kishore Mugada
Sardar Vallabhbhai National Institute of Technology Surat, Surat, India

7.1 INTRODUCTION OF ADDITIVE MANUFACTURING

Additive manufacturing (AM) is a new addition to the modern manufacturing methods in which the parts are created by the layer-by-layer deposition of material. Unlike conventional subtractive manufacturing techniques, where components are generated by removing an excess amount of material from a blank, additive manufacturing process have lots of potential in the manufacturing of components made up of metals, alloys, composites, ceramics and plastics. In these processes, the parts are being manufactured based on a three-dimensional model. In recent years, a substantial amount of research has been carried out on the various problems related to additive manufacturing techniques which enables the technology to make components of any possible size [1, 2]. These manufacturing techniques can also be used to manufacture items ranging from the simplest to the most complex geometry, many of which are designed to be commonly used in electronic and bio-medical industries [3]. In the processes, materials made of alloys, composites, plastics, bioengineering tissues and polymers are widely used as a feedstock material in powder form or feedstock rode. The feedstock material is first heated up with the help of some heat source, such as a laser in laser-based additive manufacturing, or a welding torch as in wire arc additive manufacturing processes. The heated materials are allowed to deposit on a substrate or a previous layer. The introduction of a laser as a heat source permits the exertion of great control on the intensity of melting of the materials. Therefore, laser-based additive manufacturing has been widely used for the manufacturing of most intricate components [4, 5].

The structural integrity of the AMed components is largely determined by the static and dynamic strength. Because the parts are often used for the structural applications used in automobile, bio-medical, and aerospace industries, where they are subjected to complex loading at different temperatures [6, 7]. Sometime, the loading is multiaxial and dynamic in nature,

DOI: 10.1201/9781032616025-7

which leads to the failure of the components. Residual stress and defects are two of the major challenges in the AM, which affect the structural integrity of the parts. The process parameters, atmospheric conditions and microstructure are some of the important aspects which affect the mechanical properties.

Most AM processes produces fine grains in the builds when compared with conventional manufacturing processes. As a result, the AMed parts have better static strength which enable the parts to meet the industry specifications. However, the dynamic (fatigue) strength of parts is severely affected by the defects and residual stress of the parts. Therefore, the AMed parts requires through weld inspections so that most of the defects can be studied. The most common defects in AM are the lack of fusion (LOF), insufficient bonding, tunnel defects and cavities. Sometime cracks have also been seen in the build due to residual stress. In the mechanical loading on the AMed builds, the cracks initiate from the cavities, pores, and inclusions because these imperfections increase the stress concentration in the builds. The defects have a significant effects on crack propagation in the fatigue loading. The AM imperfections also promote stress corrosion on the builds and hence have an impact upon the life of the components. Therefore, defects analysis in AM parts are needed so that defects can be characterized, categorized, and ranked based on the degree of their influence on fatigue life. There are different types of weld inspection methods in the AMed parts which can be categorized as either destructive or non-destructive techniques. The details of defects analysis have been explained in the subsequent sections.

7.1.1 Solid-state additive manufacturing

The solid-state additive manufacturing techniques are based on mechanical deformation where the mechanical energy is utilized for the destruction of the oxide layer following severe plastic deformations, which leads to the formation of metallurgical bonding. They are based on ultrasonic vibration, the supersonic speed of metal powders and friction.

The most common types of solid-state additive manufacturing are

(a) Friction stir additive manufacturing (FSAM)
(b) Friction stir surfacing additive manufacturing
(c) Friction stir extrusion
(d) Ultrasonic assisted additive manufacturing
(e) Cold spray additive manufacturing

In the following section, different types of solid-state manufacturing processes are discussed.

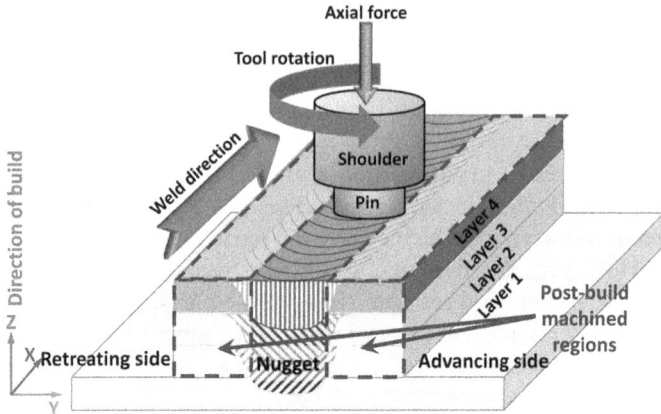

Figure 7.1 Friction stir additive manufacturing [10].

7.1.2 Friction stir additive manufacturing

The friction stir additive manufacturing (FSAM) process was patented by White [8] in 1999. It works on the principle of friction stir welding (FSW) and a solid-state AM technology where the multiple plates/sheet are friction stir welded one on top of another (see Figure 7.1). Airbus and Boeing were the first to demonstrate it, and it was suggested as a potential solid-state manufacturing option with the ability to achieve high builds at high rates with less material waste. Similar to FSW, FSAM employs a non-consumable rotating tool that is inserted into the overlapping metal sheets and then moved along the joint line to produce a weld. The material becomes plasticized by the heat produced by friction between the tool and workpiece contact surfaces, and it then flows axially and circumferentially around the revolving tool to form the weld. The process is repeated until a desired build height is obtained. The FSAM has been used for a stiffener-on-skin that is typically machined from a huge block of metal and has a very low buy-to-fly ratio (weight of raw materials/weight of the finished product) [9]. In the aerospace and aviation industries, where stiffeners/stringers can be produced by a number of methods, multilayered stiffened panels may find use.

7.1.3 Friction stir surfacing additive manufacturing (FSSAM)

In friction stir surfacing additive manufacturing (FSSAM) processes, the depositions of materials take place into two phases: initial heating, and diffusion and deposition (Figure 7.2). In the initial phase, the tool is lowered towards the substrate while maintaining a very small gap between the bottom surface of the tool and the substrate [11, 12]. After setting a narrow gap between the feed rode and the base plate, the rode is allowed to rotate

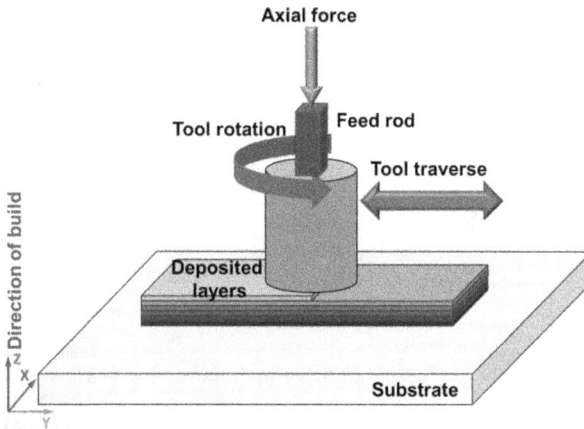

Figure 7.2 Schematic representation of friction stir surfacing additive manufacturing [10].

about its axis and given downward feed at a very slow feed rate. As it made contact with base plate with an axial force, the generation of frictional heat and plastic deformation take place at the rod–base plate interface. The initial heating takes place due to frictional heating and deformation. The initial heating in FSSAM takes place in two stages. In the first stage, frictional heat is generated when the rode makes its initial contact with the base plate. The rate of heating is slow due to lower surface contact. However, after some time the feed rode becomes softened under the effect of initial heating. Once the rode material becomes heated to a sufficient level, it undergoes an extrusion action due to the subsequent cycle of downward feed and axial force. This is the second stage of heating, where the rate of heating is significantly higher. The extrusion of material increases the frictional heat at the interface and fills the gap between the rode and the base plate. Later in the second phase of the processes, the extruded material becomes severely compressed and stirred by the rotating feed rode under the application of axial force. This leads to the diffusion of the material from the feed rod to the base plate. At this stage the traverse motion is also given to the base plate which leads the uniform deposition of feed rode material into the base plate.

7.1.4 Friction stir extrusion additive manufacturing

In FSEAM, feed material is introduced through a hollow shoulder. The feed material can be in the shape of a rod or a powder. The filler material softens as the shoulder rotates, creating heat from friction that causes it to adhere to the substrate surface (Figure 7.3). This is a relatively novel technology that was motivated by the friction stir additive manufacturing (FSAM) technique for sheet lamination AM, in which the rotating shoulder contains a

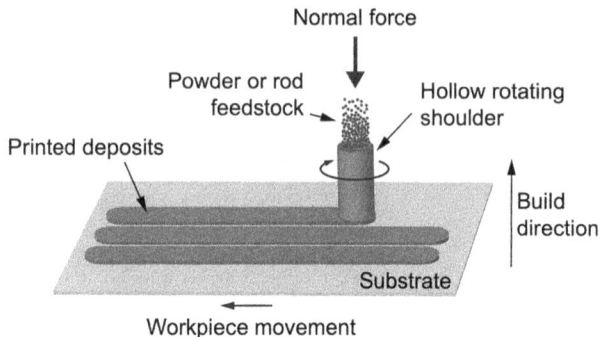

Figure 7.3 Friction stir extrusion additive manufacturing [13].

pin that is inserted into stacked metal sheets as it passes the surface rather than a hollow core. According to reports, the material undergoes strain at a rate of 1 s^{-1} to 100 s^{-1}, which is several orders of magnitude less than other solid-state AM processes. Even though the process results in high temperatures, a regulated environment is not necessary because the heating is only local. This also makes it possible to implement onsite repair with FSEAM. A further benefit of FSEAM is that it allows for the deposition of several materials, making it possible to produce bulk, homogenous metal-matrix composites like Al-SiC.

The main factor affecting the bonding, microstructure, and mechanical properties is the material flow in the friction zone. The processing factors that govern the material flow and heat generation in the work area, also known as the "stir zone," are the tool parameters (geometry, rotational speed, traverse velocity, and offset distance), filler feed rate, substrate characteristics, and cooling media. The way heat is produced and exchanged with the environment is said to be the reason why the temperature profile in the SZ of an FSEAM differs from that of an FSAM. There isn't many systematic research on how the processing parameters and final attributes relate because this is a relatively new procedure.

7.1.5 Ultrasonic assisted additive manufacturing

In ultrasonic assisted additive manufacturing (UAAM), a stack of identical or different metal foils is bonded together by high-frequency (usually 20 kHz) ultrasonic vibrations while also applying a normal force. The UAAM techniques has been schematically shown in Figure 7.4. Periodic machining procedures are then performed to give the part its final shape. Unlike the other methods, UAAM is a sheet lamination process. Foils with thicknesses in the range of a few hundred microns serve as the feedstock for UAAM. In the processes, a sonotrode which is allowed to vibrate perpendicular to the build direction transmits the ultrasonic vibrations to the foil surface from a

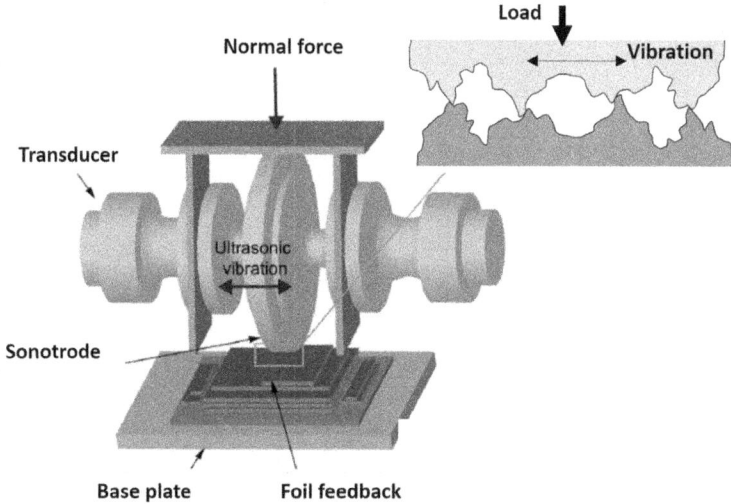

Figure 7.4 Ultrasonic assisted additive manufacturing [13].

low- (1 kW–2 kW) or high-power (up to 9 kW) transducer. Before the installation of the next foil, the sonotrode's rough surface causes microasperities on the foil surface. The processes apply deformation at a strain rate in the range of 10^4 s^{-1} and 10^5 s^{-1}, and also applies the normal force and high-frequency vibrations [13]. This leads to the collapse of the microasperities and the extreme plastic deformation at the interface, which causes bonding.

The porosity, the existence or absence of continuous metallurgical bonding, and the oxide breakdown at the interface are the three main factors that determine the strength of an ultrasonically made component, particularly in the build direction. By keeping the seams from aligning, overlapping and staggering the foils can reduce or avoid void formation at the foil junctions and improve the mechanical qualities.

7.1.6 Cold spray additive manufacturing

The cold spray technique was initially used as a coating method in the 1980s. Recently, the cold spray has emerged as a new additive manufacturing technology. Cold spray additive manufacturing (CSAM), which depends on supersonic impact of powder particles into a substrate, is another type of solid-state additive manufacturing technology that produces bonding by mechanical deformation (Figure 7.5). A heated propulsive gas (25°C to 1000°C), usually nitrogen, helium, or air, accelerates a powder feedstock carried by a cold compressed gas to supersonic velocity towards a substrate. The powder particles' kinetic energy is primarily transferred into their plastic deformation upon impact in this line-of-sight process. The breaking of

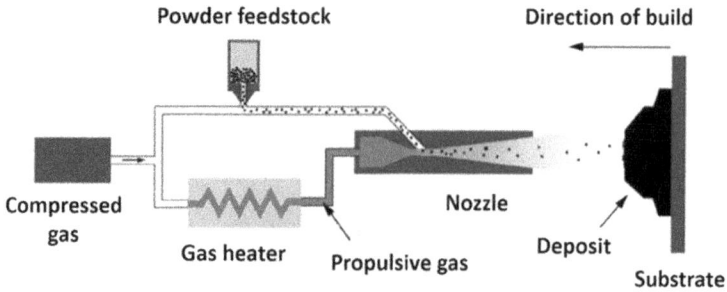

Figure 7.5 Cold spray additive manufacturing [13].

the surface oxide layers and severe plastic deformation caused by the collision facilitate bonding by causing metal jetting, which is characterized by a strong outward material flow. Bonding has also been observed to be aided by mechanical interlocking between the powder particle and the impact crater.

In CSAM, the speed of the powder particles must exceed a critical impact velocity in order to ensure strong bonding and high efficiency deposition. The density, ductility, strength, shock and thermal properties, as well as the preheating temperature, of the powder and substrate, and also the impact parameters (powder size, impact angle, substrate and particle starting temperatures), determine the critical velocity for a material; For instance, for a specific powder-substrate configuration, the critical velocities for aluminum, nickel, copper, and zinc are in the following sequence, from highest (800 m/s) to lowest (500 m/s): Al > Ni > Cu > Zn.

7.2 COMMON DEFECTS IN SOLID-STATE ADDITIVE MANUFACTURING

Various types of defects have been reported in solid-state AM of similar and dissimilar materials. The majority of these defects in friction stir-based AM occurs due to the improper selection of the process parameters, tool design, weld environment and fixture. There are different types of AM defects and their formation possibility depends on various process conditions. These defects are lack of fill, lack of fusion, kissing bond, root flaws, nugget collapse, wormhole, lack of penetration, void etc. These defects are primarily caused because of improper material flow and the insufficient or excessive amount of the heat input in FSAM. Literature [14, 15] have classified weld defects in the various categories depending on geometrical shape, locations and heat input condition. According to geometrical shape, defects can be line defects, point defects, volumetric defects etc. According to location perspective, defects can be categorized into internal defects and surface defects. Similarly, depending on the heat input condition, weld defects may arise due to cold condition, excessive hot condition, and faulty tool design. The following paragraph discusses the weld defects from the perspective of location, i.e. internal and surface defects.

7.2.1 Internal defects

This defects cannot be visible from the weld surface. These defects can easily be identified by using either destructive or non-destructive tests.

7.2.1.1 Voids and tunnel defect

These defects are volumetric defects which are mostly found at the SZ. Both tunnel and void defect can be identified and quantified through the use of non-destructive tests. Tunnel defect is found continuous along the weld span whereas void is discontinuous, and can be found at specific locations. Both these defects are aligned along the welding direction, where there is an absence of material. Specifically, both these defects are found along the AS of the weld zone due to inadequate pressure and heat. In FSAM of dissimilar materials such as aluminum-steel, the formation of voids and tunnel occur when there is large amount of steel flashes with the Al substrate (Figure 7.6). Steel flash produced an inhomogeneous material flow by hindering the material to fill along the narrow cavities available in the steel particles [16]. These defects can be controlled by increasing the value of PD, α and reducing the value of v.

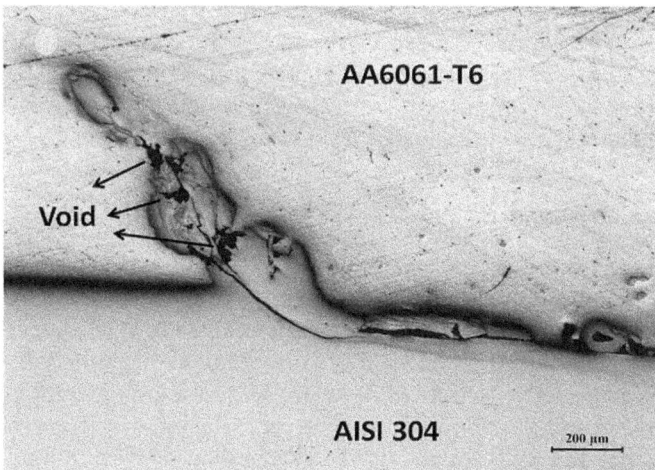

Figure 7.6 Voids in FSAM of Aluminum and steel.

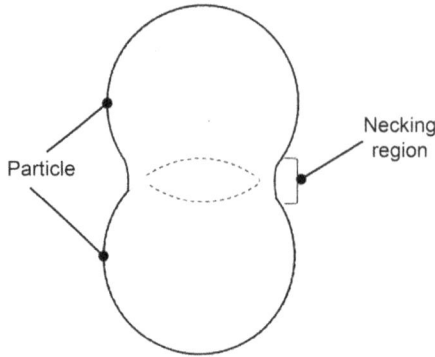

Figure 7.7 Schematic of necking regions where the pore forms [17].

7.2.1.2 Sintering porosity

This type of porosity is mostly seen in sintering-based additive manufacturing where the thermal processes are applied on the powder form of material. The heat and pressure in sintering-based AM led to diffusion, recrystallization and grain growth. In these processes various mechanisms have been seen and they are mass transfer, surface diffusion, evaporation condensation, grain–boundary diffusion, plastic flow, viscous flow and lattice diffusion. These mechanisms establish bridges among the powder grains. In sintering-based AM, the metals in powdered form are converted into the material in bulk form. The strength of bulks AM being lies due to the metallurgical bonds among the neighboring particles (Figure 7.7). The pores in the sintered components have been seen between neighboring particles, and are believed to form during the initial and intermediate stages of the sintering processes. In these processes, the formation of pores starts as an interconnected channel along the boundary of three particles. Later, due to non-uniform shrinkages among the metallic powders, the final shape of pore emerges, which separates the particles.

7.2.1.3 Kissing bond

This is an internal defect which cannot be detected by the non-destructive test. A kissing bond is meant to refer to partial or weak bonding between two layers, as shown in Figure 7.8. These defects are formed when the oxide layer of the materials to be joined could not become fragmented during AM. Oxides of pure metal, such as oxide of Al, possess a high melting point and also different mechanical properties than the base Al. So, the presence of the oxide layer in the weld zone forms weak bonding between the two layers. In addition, under high heat input and material flow, the materials in the processed zone undergoes a sliding action with the un-deformed material

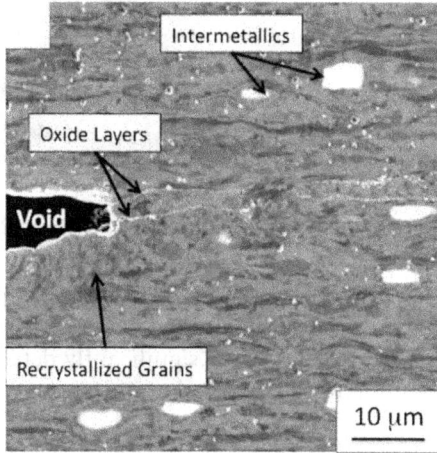

Figure 7.8 Defects (kissing bond) in ultrasonic AM build due to the presence of oxide [18].

which again results in the formation of the kissing bond. This defect can be reduced by increasing the value material deformation rate and pressure.

7.2.1.4 Lack of fusion

Low heat input is the main reason behind the formation of this defect. Lack of fusion mostly occurs in the butt weld configuration where the two faying surfaces could not fuse with each other. This causes significant reduction in the build strength. By increasing the value of PD and ω and D_s in FSAM, this defect can be avoided.

7.2.1.5 Root flaws

These mostly occur when the workpiece sticks with the backing plate or anvil in FSAM. The sticking of the workpiece occurs when the length of the tool pin when compared with the thickness of the materials being welded. In addition, the surface state of the backing plate also affects the sticking of the workpiece material. When the backing plate does not have a smooth flat surface, then the workpiece takes on the impression of the surface irregularities of the backing plate. These solid-state AM defects can be eliminated by using shorter pin lengths and the proper selection of the backing plate or anvil.

7.2.1.6 Lack of penetration

Lack of penetration (LOP) mostly occurs beneath the SZ of the friction-based AM. This defect refers to regions where metallurgical diffusion does

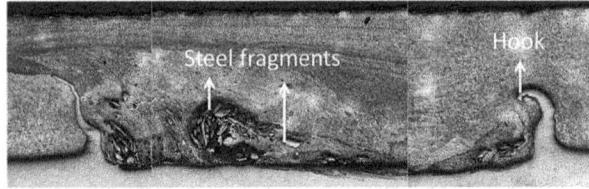

Figure 7.9 Hooking defects in FSAM of aluminum and steel [23].

not occur, and is believed to be caused because of shorter pin height as compared to the sheet thickness. LOP is initiated with cracks under the effect of static and dynamic loading. This defect significantly deteriorates the tensile and fatigue strength of the weld build and non-destructive tests can be used to identify this defect.

7.2.1.7 Hooking defects

Hooking defects mostly occurs in the builds along advancing side (AS) and retreating side (RS) at TMAZ (Figure 7.9). This defect occurs when the tool penetrates the lower sheet in lap welding. Penetration of the tool deforms the lower sheet and pulls out the deformed material. The pulled-out deformed material becomes accumulated around the pin or the SZ, which causes a distorted interface on the flat interface. In FSAM of dissimilar materials such as Al and steel, these distorted regions surrounding the pin produces an unbounded region, a fault which is known as a hooking defect. The accumulation of steel fragments in the Al matrix hinders the material flow by stopping the material filling among the cavities of steel fragments, thereby leading to the formation of voids. The hooking defect reduces the sheet thickness of the upper substrate, leading to a reduction in the load-bearing area of the weld zone against the static loading and dynamic loading. Thus, under fatigue loading, many authors have reported the initiation of cracks from hooking defects [19]. Many authors [20–22] have extensively studied the effect of pin shape, pin depth, ω and v on the hooking defects. They reported that the formation of hooking defects can be avoided by optimizing pin depth, ω and v.

7.2.2 Surface defects

These defects are formed on the weld surface which are clearly visible. The details of surface defects have been discussed in the following sections.

7.2.2.1 Groove defects

Figure 7.10 shows groove defects formed on a FSAMed sample. This defect generally occurs along the AS of the weld surface. Low heat input and

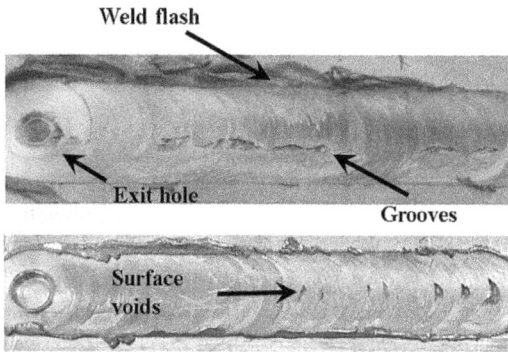

Figure 7.10 Surface defects [15].

insufficient *PD* are the major reasons behind the formation of this type defect. This defect occurs when the material from RS could not reach the AS. This defect can be avoided by optimizing *PD* and α.

7.2.2.2 Weld flash

This defect is produced when welds are fabricated with excessive high heat input and higher plunge force. High heat is generated with extremely lower values of v and higher values of ω. The high heat softens the materials and the larger value of the plunge force expels the material from the weld zone. The expelled material is then deposited around the weld zone. This defect reduced the thickness of the weld zone when compared with the thickness of the base material. Weld flash formation can be avoided by providing a α to the tool and by reducing the value of ω and *PD*.

7.2.2.3 Surface galling

This defect mostly occurs during FSAW of softer material such as Al and magnesium (Mg). During FSAM, under the effects of high heat input and severe plastic deformation, weld material becomes stuck to the FSW tool. This leads to the formation of surface defects. It can be avoided by reducing the value of ω and *PD*.

7.3 METHODS FOR THE ASSESSMENTS OF BUILT QUALITY

There are different types of methods for the assessments of the defects in additive manufacturing builds. They can be categorized in two main types: Destructive and non-destructive.

7.3.1 Destructive tests

7.3.1.1 Optical and scanning electron microcopy

Microscopy, based on light optical and electron processes, are the destructive techniques for the studying the fine details of defects in AM builds. The SEM technique provides a high resolution of an image in microscopic scale. It has been used to witness size, shape and location of defects on a cutting plane. Both techniques enable to give two-dimensional (2D) projections of defect at a particular location. The techniques can be used to capture 2D images at different locations of a sample and later they can be stitched. Stitching technique has been used in many literatures to detect the defects through the AM builds length. The SEM can also be used to generate a 3D view of defects by stitching 2D images of defects at different cutting plans. However, this technique requires longer time and rigorous work related to cutting, grinding and polishing of samples at different planes. The SEM and the OM techniques are the least expensive methods for defects characterization in AM builds.

7.3.2 Non-destructive tests

(a) Die-penetration tests
(b) X-ray micro computed tomography
(c) Infrared imaging defect detection
(d) Gas pycnometry

7.3.2.1 Die-penetration tests

This is one of the oldest non-destructive methods for the defect identifications where the defects are identified due to capillary action of die (penetrant) applied over the surface. In this technique colored die (penetrant) is applied over the surface of sample; this infiltrates and settles on the defective surface (Figure 7.11). Later, the excess residue penetrate is removed. In

(a)   (b)

Figure 7.11 Die-penetration results of (a) rocket gas injector (b) POGO-z bazel [25].

order to visualize the defects properly, a reactant (developer) is applied on the surface which reacts with the penetrant in the presence of light. After the reaction between the reactant and dies, defects and the reacted penetrant can easily be seen. The accuracy of the defect detection in the technique is largely dependent on the fluorescent agents, the types of defects and the capillary action of penetrant. The die penetration is a less expensive non-destructive testing (NTD method) for the quality assessment of the AM parts. Therefore, it is now being used in various aviation industries and R&D centers for the quality assessment of rocket engines, baffles etc. which are generally manufactured using selective laser melting or electron beam-based AM techniques [24, 25]. However, this technique cannot be employed for the detection of internal pores in AMed parts. In addition, it is also impossible to test very rough AM parts using the die-penetration technique. In such instances, post-processing, such as polishing, micro-machining etc. of the surfaces of AM parts, is required for defect identification.

7.3.2.2 X-ray micro computed tomography

The X-ray micro computed tomography (MCT) technique uses X-rays to scan the sample at different angles. The procedure of this technique is shown in Figure 7.12 where the sample is mounted on a rotated table between the X-ray source and a detector. When the X-ray beam falls on the sample, it passes through the sample and produces gray-scale image on the projector or the X-ray beam get annulated. The quality of projected image determines the accuracy of the defects. In the tomography technique the resolution of projection is determined by the magnification factor (Voxel size) of the object which is decided by the relative position of geometry of source.

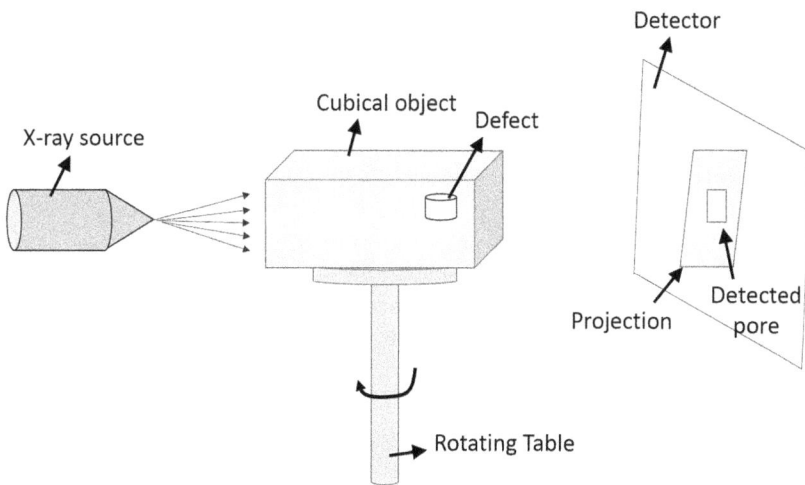

Figure 7.12 Schematic Diagram of X-ray micro-tomography.

During the test large number of images (projections) of the images are collected at different angular positions of the object by rotating it 360°. Later the three-dimensional (3D) view of the object is reconstructed by using back projections of all images. The volume of the 3D model consists of 3D pixels (which is known as a voxel). This consists of a grey value between 0 and 65535 proportional to relative attenuation of the X-ray by the material.

The fatigue performance of AMed parts are affected from the distributions of defects and its geometrical features which includes the size, and morphology. In much of the literature, the defects have been distributed non-uniformly in the AM built. However, few studies have been reported that the major fraction of defects get localized closer to the surface of AM parts than the core. High pressure and a low thermal gradient are the likely explanations for the lower number of defects at the core of AM parts. Low axial pressure, and a high temperature gradient, has been noticed at the edge of the AM built during the process, which leads to the existence of non-uniform consolidations of materials in the build. In addition, the high temperature gradient led to the thermal expansion and contraction of materials at the edge of the build during the heating and cooling cycle of AM. This also leads to the development of residual stress in the built and hence cracks, and the poor surface finish at the edge of the AM build.

The internal defects in AMed parts are generally access by using the non-destructive technique, meaning that they are relatively more costly than parts that are manufactured from other processes. The most widely used non-destructive techniques are the Archimedes method, gas pycnometry and X-ray micro-computed tomography (X-ray MCT). The first two methods are rarely used in the defects analysis in AM parts as they cannot give information regarding the exact locations, size and distributions of internal defects. Both Archimedes and gas pycnometry can provide qualitative information about the defects such as high and low density represents fewer and higher porosity in the build. The X-ray MCT can provide the three-dimensional topography of internal defects. The technique can detect the size, diameter, projected area, and locations along different directions. The X-ray MCT has also been utilized for the predictions of cracks in AM in-situ, both during and after the processes. The accuracy of X-ray MCT results is dependent on the voxel size and X-ray penetration into the workpiece. Increasing the size of the object requires low scanning speed and smaller voxel size because the resolution of X-ray image will degrade. The defects cannot be measured accurately at a higher scan speed of a larger specimen because the level of X-ray penetration decreases with size of the objects. Therefore, the larger the size of the specimen, the longer the duration of X-ray exposure for the accurate detection of defects. The density of the AM is another important factor which affects X-ray penetration. Smaller voxel size is required to scan the AM parts made of materials with higher densities. The selected voxel size during X-ray MCT should be smaller than the size of the pores in AM parts. Therefore, the accuracy and reliability of X-ray MCT is dependent

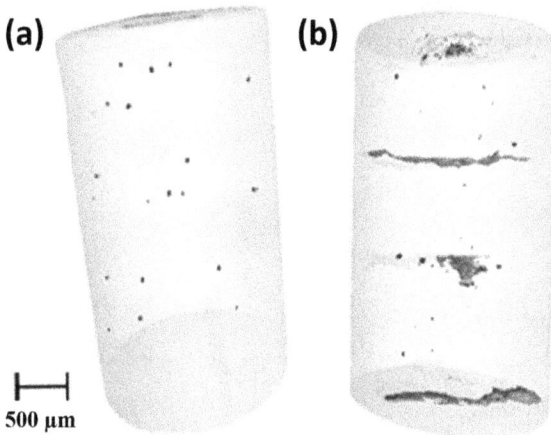

Figure 7.13 SRμT detected pores in AM build [26].

on the size and density of the materials. Researchers have reported a critical number of voxel number 27 in volume to detect the internal defects accurately. Smaller than the given critical voxel number leads to the presence of several voxel noises in the scanned image and hence defects cannot be scanned accurately.

X-ray MCT may be unable to predict the fine details of defects (Figure 7.13). Sometimes the very small size of the internal defects cannot be predicted by using X-ray MCT. Such fine details about the internal defects can be studied by Synchrotron Radiation Micro-Tomography (SRμT). SRμT uses a high-density X-ray source to scan the samples. This technique possesses a high signal-to-noise ratio and can detect the sub-micron size of internal defects. The SRμT is a most advanced defects characterization technique which provides the best resolutions of images as it uses high-energy density X-ray beam. The technique has been used by a few researchers to characterize the defects, both online and offline, in AM and welding processes. However, the SRμT is very expensive and it requires a very small sample size. Therefore, this technique requires a longer time than the X-ray micro CT for the characterizations of defects.

The X-ray MCT may be unable to detect the internal defects such as lack of diffusion, sharp corners of cracks and the kissing bond in the AM build. In addition, defects analysis with X-ray MCT is very costly. Recently analytical, image and signal processing techniques have been utilized by few research groups for the monitoring of defects both offline as well as online in welding and AM processes. The details of these modern techniques have been described in the next section.

(c) Infrared imaging defect detection

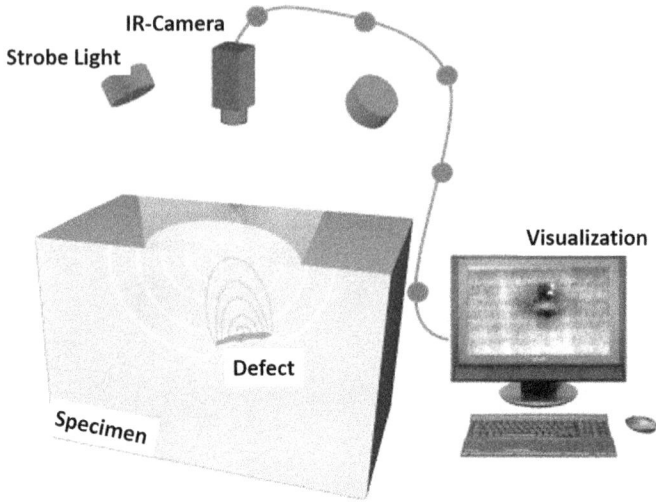

Figure 7.14 Working principle of thermography [29].

In this technique the shape and contour of the defects are identified through the intensity of thermal radiation which is allowed to pass through the samples [27]. Figure 7.14 shows the working mechanisms and important parts of this technique. The defects and the sample produces changes in the intensity of thermal radiations which are being used for the detection of defects. The technique is able to display the defects in infrared images. The essential presence of defects in the sample affects the heat conduction and hence the temperature field in the sample. Later the infrared sensors are used to detect the temperature field which helps in the predication of defects. Therefore, the technique is able to detect the location of defects in the sample. However, it is unable to quantify the size and shape of defects. One research study [28] detected the defects in the electron beam-based AM processes by using a long-wave infrared camera. The predicated defects were compared with the experimental observations. Defects were identified on samples where the difference in the temperature field was more than 1% between layers. The infrared camera can also be used for the defect detection in real-time AM. Few literatures suggested that the technique can detect more unfused defects of size lager than 500 μm. However, the infrared camera cannot detect defects of micron size.

7.3.2.3 Artificial Intelligent approach

1. Image processing approach

   The build quality in solid-state AM can also be studied by using the image processing approach after collecting images of the build layer by layer. The technique is also known as machine vison system which involves image acquisition, processing and decision. The system is capable of providing the features of AM build directly in 2D and 3D planes.

Therefore, the image processing techniques are providing reliable build quality with better repeatability and accuracy than other techniques.

In solid-state metal AM, the material is getting deposited in the build layer by layer is solid-state. The deposition of material in multiple layers in the build decide the build geometry and AM part. The build quality of the AM is largely dependent on the substrate or previously deposited layer on the build. The surface of the substrate or already deposited layer can be monitored by using image processing techniques. Later the features of surface such as crack, grooves, and other surface defects are getting detected. In addition to that machine vision system also predicts the width and thickness of the AM builds. Based on the presence of surface defects, and size of the builds, the process parameters of AM can be decided to eliminate the defects.

The following steps to be followed for image processing:

(a) **Application or processed zone**
(b) **Image acquisition:** The accuracy of image processing techniques is primarily dependent on resolution, and contrast. Field of view is an another important requirement in the field of real-time in-situ image processing of an additive manufacturing.
(c) **Image enhancement:** The images of AM builds contain lots of noise content which has to be removed by increasing the edges or contrast in frequency or spatial domain. Later the enhanced images are processed which included image segmentation and feature extraction. The spatial domain include uses of filter for noise reduction by using convolution operation. Sometime, noise reduction in spatial domain also take place through the modification of the histogram of the image. In the frequency domain, the image enhancement is carried out by using Fourier transformation which involves element to element multiplications. It reduces the computational time. Later the inverse Fourier transformation is applied on the enhanced image to obtain a processed or enhanced image. The brief descriptions about of the operations related to spatial and frequency domains are given below:
(d) **Image segmentation:** In this stage of image processing, regions of the image are segmented by using segmentation tools. It is carried out by using tools such as edge, line, point and region-based detection-based segmentations.
(e) **Features extraction:** The features of an image are extracted from an image by the edge detection technique. It has been applied for the applications in various manufacturing processes such as machining, welding and additive manufacturing. Edge can be defined as the sudden change in the gray-level intensity from its neighbourhood. Change in gray-level intensity could be along a line or curve. The feature detections are carried out by three steps: (a) Image smoothening (b) Detection of edge points (c) Localization of edge. The noise from the

image is becoming eliminated in step one followed the thresholding operation and localization in the second and third steps, respectively. Later based on the threshold value of features, defects are classified which is termed as classification/measurement.

7.3.2.4 Signal processing approach

Signal is a function that conveys information about a physical phenomenon as a sensor output. For example, temperature profile in the welding zone monitored using the thermocouple (sensor) is a signal. The signal can use either continuous or discrete values. A signal can be defined as a continuous signal if it has values of finite or infinite range, example an electric current signal drawn from an electrical device. However, if a signal has values of finite set can be defined discrete signal, example average of grades of a class.

If the signal is continuous in both time and value, it is called an analog signal, as shown in Figure 7.15(a). If the signal has discrete time and continuous values, it is a sampled signal, as shown in Figure 7.15(b). If the signal has discrete time and discrete values, it is termed a digital signal, as shown in Figure 7.15(c). If the signal has continuous time and discrete values, it is termed a quantized signal, as shown in Figure 7.15(d). Signals which are generated by multiple sensors are called multi-channel signals.

Signal processing is an operation in which the characteristics of a signal, such as amplitude, shape, phase, frequency etc., undergo a change. The typical interpretation of quality using the sensor input data, signal processing is shown in Figure 7.16. The unwanted signal that is interfering with the main signal is termed as noise and can be filtered, however noise is also a signal but unwanted.

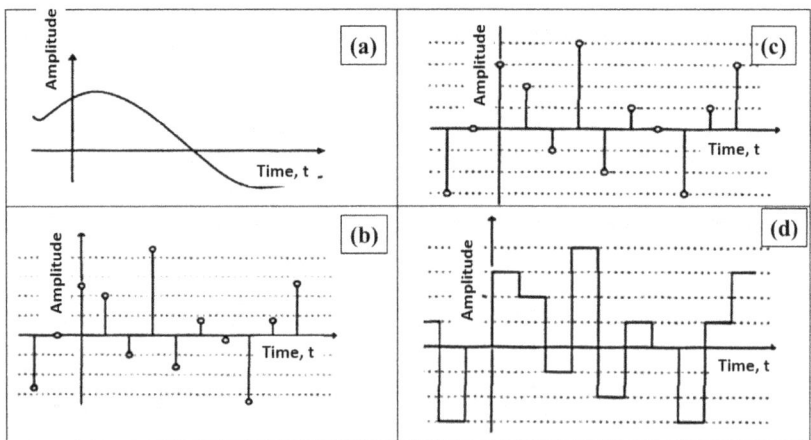

Figure 7.15 (a) Analog signal, (b) sampled signal, (c) digital signal, (d) quantized signal.

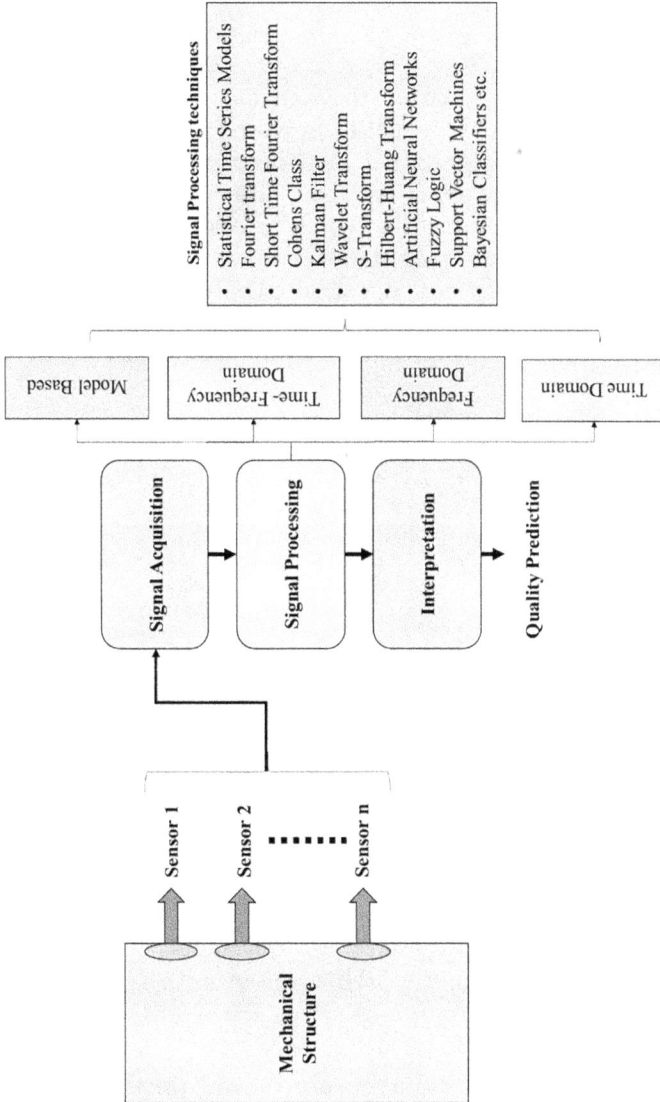

Figure 7.16 Sensing and signal processing in the industrial applications.

Signal processing methods are classified in to four groups: time domain, frequency domain methods, time-frequency domain methods and model-based methods.

7.3.2.4.1 Time domain methods

In this method the signal is processed in only the time domain. The frequency domain is not applied on the signal during the processes. The energy and magnitude of the signal is estimated in the time domain signal processing method. Later peak-to-peak height, average, area under the curve and slope of the processed signal are estimated. Moments of statics are used on calculated value of the signals to calculate various statistical parameters such as mean, standard deviation, kurtosis etc. for diagnosis. If a discrete signal $x(n)$, *let* N and x denote the length and the mean the statistical parameters range (Max-min), root mean square value (RMS), crest factor and kurtosis can be calculated by using following Equations 7.1–7.5.

$$\text{Peak} - \text{to} - \text{valley} = \max\big(x(n)\big) - \min\big(x(n)\big) \tag{7.1}$$

$$\text{RMS} = \sqrt{\frac{1}{N}\sum_{i=1}^{i=N} 1\big(x(i) - \bar{x}\big)^2} \tag{7.2}$$

$$\text{Crest Factor} = \frac{\text{Peak} - \text{to} - \text{valley}}{RMS} \tag{7.3}$$

$$\text{Kurtosis}\big(\text{normalized 4th moment}\big) = \frac{\frac{1}{N}\sum_{i}^{N}\big(x(i) - \bar{x}\big)^4}{\text{RMS}^4} \tag{7.4}$$

Probability density function

$$p\big(x \le x(n)x + \Delta x\big) = \frac{\text{No. of } x(n) \text{ between } x \text{ and}\big(x + \Delta x\big)}{N} \tag{7.5}$$

A typical flaw with time domain approaches, with the exception of auto-correlation, is that they don't reveal information concerning periodicities. RMS is a measure of signal energy that, while it doesn't by itself provide further diagnostic information, can indicate flaws or be linked to certain process characteristics.

7.3.2.4.2 Short-time signal processing

Sometimes, it is necessary to deal with non-stationary signals whose characteristics, including energy and mean, slowly vary over time. This results in a number of "short-time" processing techniques wherein brief portions of the signal are separated and treated as though they were brief portions of a sustained signal with fixed characteristics. Periodically, this is done as often as needed. These brief sections frequently cross over. A single number or a collection of numbers may emerge from the processing of each frame. As a result, this processing generates a new time-dependent sequence that can represent the signal.

7.3.2.5 Synchronized averaging

In order to improve the signal-to-noise ratio of a periodic signal, this algorithm is widely employed as a pre-processor. The method entails ensemble averaging a series of signal segments, each lasting one fundamental cycle, such as one rotation or one-part cycle shown in Figure 7.17. In order to ensure that samples are obtained at the same location each cycle, it is believed that a sensor, such as an encoder, is used to clock the sampling. A fixed number of samples will be produced by each cycle, which we refer to as $x_i(n)$, where i is the cycle number and n is the sample number. Then, one may compute the point-by-point average across M cycles as Equation 7.6.

$$x_{ave}(n) = \frac{1}{M} \sum_{i=1}^{M} x_i(n) \tag{7.6}$$

7.3.2.5.1 Frequency domain methods

The analysis of the signals in the frequency domain is very popular. Fast Fourier transform (FFT) algorithms are used in frequency domain-based signal processing methods. The discrete Fourier transform (DFT) algorithm

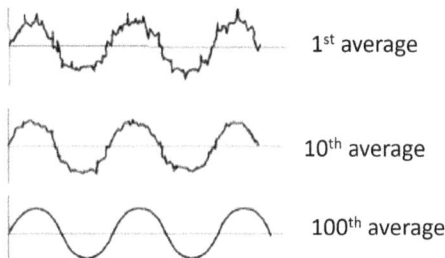

Figure 7.17 Synchronized averaging.

has been extensively applied to signal processing to identify the periodic phenomenon and determine their strength. The distribution of energy over frequency is studied in power spectra. The spectra are plotted with multiple values of rotational speed of machine whose value could be an integer or fraction because the variation in the magnitude of energy (signal) is proportional to the rotating frequency of the machine. The processing of the signals often require a filtering operation which is accomplished by applying different weight to frequencies. The common types of filters are highpass, lowpass, and bandpass/stop filters.

7.3.2.5.2 Time–frequency methods

This method is used for the processing of the signals which are nonstationary and their distribution of energy over frequency changes over time. The traditional spectral analysis techniques inadequate to process the non-stationary signals. These types of signal is often seen in additive manufacturing processes such as magnitude of spindle force and spindle motor power consumption. The magnitude of these signal changes over time. The models used for the time-frequency methods have been discussed below:

1. **Statistical Time Series Models:** The statistical time series (TS) models are the oldest type of the signal processing technique. These models can be classified into linear and non-linear statistical TS models. Linear TS statistical models are simple to implement and model linear systems with minimum error in diagnosis of defects in AM processes. The most commonly used statistical models are autoregressive vector (ARV), autoregressive model with exogenous inputs (ARX), and autoregressive integrated moving average (ARIMA). In addition, autoregressive (AR) model, moving-average (MA) model, autoregressive moving average (ARMA) model are also come under the category of the statistical TS models.
2. **Fast Fourier Transform:** The Fast Fourier Transform (FFT) algorithm in commonly used signal processing techniques for defect detection in the additive manufacturing. It has been applied on signals which are in simple harmonic motion for the analysis of stationary signals. The FFT converts continuous time series signal into a frequency domain which can be used for the defect detection in the AM processes.
3. **Cohen's Class:** Cohen's class (CC) distribution is a method to estimate the energy of time-varying systems. Different types of time–frequency distributions such as Wigner–Ville Distribution (WVD) and the Choi–Williams distribution (CWD) [24] have been proposed, but [26] shows that all time–frequency distributions can be generalized by a unified formulation in the following form:

$$D(t,f) = \frac{1}{4\pi^2} \int\limits_{-\infty}^{\infty}\int\limits_{-\infty}^{\infty}\int\limits_{-\infty}^{\infty} e^{-t\theta j - j\tau f + j\theta t} \times \varphi(\theta,\tau) x * \left(t - \frac{\tau}{2}\right) d\theta dt d\tau \qquad (7.7)$$

where x is the input signal value, x^* is the complex conjugate of x, $\varphi(\theta, \tau)$ is the kernel function or distribution (e.g. WVD), θ and t represent the frequency index and time, respectively, centered at time τ.

4. **Kalman filter:** The ideal recursive data processing algorithm known as the Kalman filter (KF), commonly referred to as the linear quadratic estimator, is capable of estimating linear dynamic systems. KF requires previous knowledge of both the system errors and measurement noise in order to estimate the linear system. In actual use, the procedure and the noise statistics are both imperfectly known. KF cannot be employed in practical structural engineering applications because of its nonlinear behaviour.

7.3.2.6 Wavelet transform

This is a relatively new technique for signal processing which uses time and scale window functions to represent the signal's time-frequency spectrum. Wavelet transform (WT) and its enhanced version, wavelet packet transform (WPT), have recently risen to the top of the list of signal processing methods. The WT is calculated using a set of discrete-time low- and high-pass filters, referred to as approximations and details, respectively. The first level's approximation is split into new detail and an approximation based on a multiresolution analysis and this is repeated. The time-frequency spectrum of the signal is represented using time and scale window functions by a relatively new signal processing method known as WT. The wavelet transforms and WPT, an improved version, have recently ascended to the top of the list of signal processing techniques. Discrete-time low- and high-pass filters, referred to as approximations and details, respectively, are used to calculate the WT. An approximation based on a multiresolution analysis and additional detail make up the first level's approximation.

7.4 SENSOR DATA-BASED METHODS

7.4.1 Thermal process-monitoring methods

Temperature in the solid-state additive manufacturing largely affects the fusion of layers during the processes. Low temperature in an AM leads to formations of improper bonding among the layers and hence defects. In

AM, the defective area is monitored through an analysis of the temperature profile. Pyrometry and infrared thermography are the two widely used NDT thermal techniques for the prediction of defects. Both sensors are used to collect the temperature gradients in the builds during the AM processes. The infrared camera provides the two-dimensional temperature gradient profile of the surface area, whereas the pyrometer provides the temperature data of the AM at a particular location. Later the temperature data are processed and obtained all the features of the signal which helps in to analyzing the changes in the thermal profiles with the change in the process parameters of AM. However, the accuracy of the predictions in both of the techniques are highly dependent on emissivity, the reflection of lights, and material flow, and burr during the AM. Such problems have been seen significantly in pyrometry, which has been tackled through the installation of high-speed cameras. The use of both sensors in the AM help in acquiring the temperature gradients of the whole processed zone as well as temperature at a particular location. The collected data are essentially providing the temperature profile and gradient of materials deposition of each layer. Hence a better prediction of the build quality can be made (Figure 7.18).

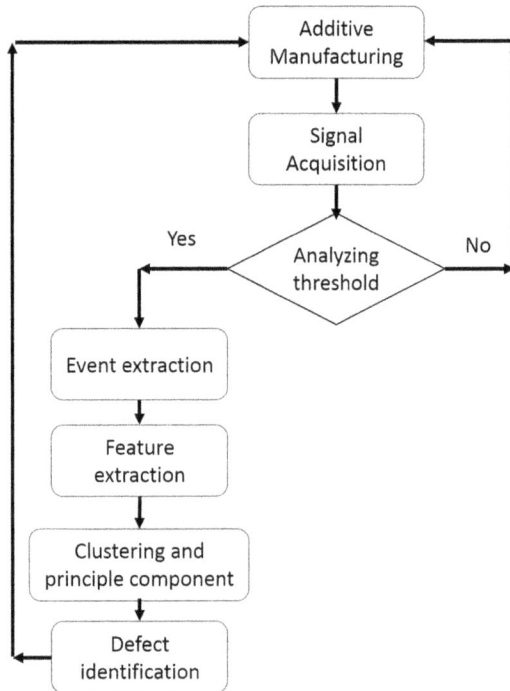

Figure 7.18 Schematic diagram representing the defect identification by using signal processing.

7.4.2 Optical process-monitoring methods

The build quality in the solid-state AM can also be studied through the use of optical measurement devices in-situ. In the technique the images of the AM builds are captured by using high speed cameras which are latter processed. The camera is generally made from charge-coupled devices (CCDs), complementary metal oxide semiconductors detectors and optical emission spectroscopy (OES). In this technique the high speed camera is generally mounted with the machine head which takes the images of the build during AM. Through the optical measurement techniques, build qualities such as the surface roughness, defects which includes cracks, infilled cavities, and size of the build can be monitored. The devices used in this technique collects the light getting reflected from the builds and hence can give sufficient information about the build surface defects. However, the technique cannot provide any information about internal defects of the build during the AM processes.

The high speed camera is available at low cost and can work in the near-infrared regime which make it suitable for the build quality inspection affordably. The OES is the latest addition in the field of built quality inspection which has been used the for the monitoring of defects and physical mechanism in the AM. Defects, such as lack-of-fusion defects, can easily be identified in the builds. The OEM is based on the emission signals of builds during the AM. It has also been used for the prediction of hardness, surface appearance and micro-structures.

Optical methods have been used widely for the quality inspection of the AM builds. However, the accuracy of the technique is largely dependent on the degree of light illumination, resolution, and capture time. In addition, in-situ build quality inspection is facing challenges, due to the difficulties with image processing techniques and the adequate speed required for the feedback loop.

7.4.3 Ultrasonic process-monitoring method

Ultrasonic techniques have been widely used for defect monitoring in welding and AM processes. The technique can be categorized into two principal types: Contact and non-contact. In both types pulsed waves are applied to the samples where mechanical energy is becoming generated which will be either absorbed or reflected from the sample (Figure 7.19). The absorption or rejection of energy is detected through a receiver which generates an electronic signal. In the testing technique, the presence of defects and its geometry is predicted based on the variation in the magnitude of the signal. In addition, ultrasonic techniques have been widely used for the investigation of elastic properties, grain size, texturing and phase transformations.

Ultrasonic techniques have been used for the quality monitoring of AM parts in-situ and ex-situ. The contact-type ultrasonic technique can provide

a

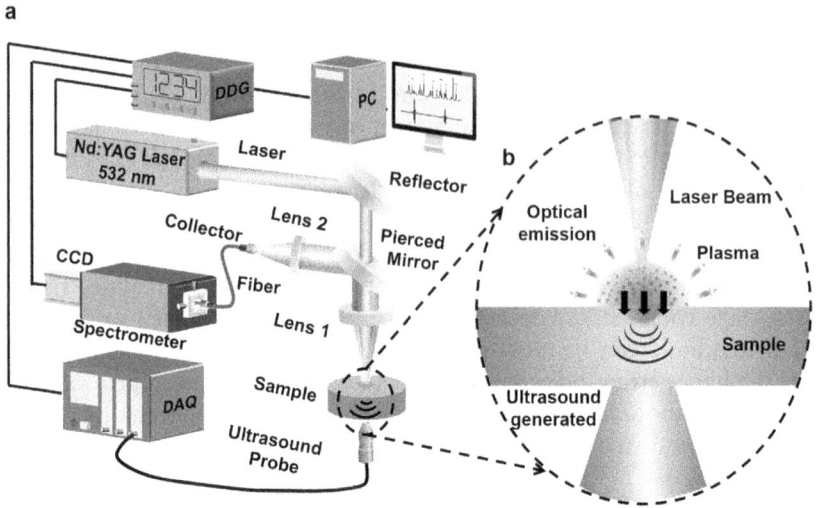

Figure 7.19 Schematic representation of Laser Unit Ultrasonic Testing method [30].

the flaw details of the surface or the subsurface of AM parts. The contact-type ultrasonic technique utilizes the variability in the wave velocity during testing the sub-surface of AM. However, after material deposition in AM layer by layer, there is a very high chance for getting noise in the collection of wave velocity which results in the error in quality prediction.

The non-contact types of ultrasonic technique are also facing some challenges related to the complexity in signal processing and the detection of AM parts alignments. However, few studies have detected the defects in AM builds through non-contact-type ultrasonic techniques which includes spatially resolved microscopy, laser ultrasonic and acoustic emission testing on an ex-situ level. In addition, the ultrasonic technique provide bulk properties of the material.

7.5 SUMMARY

Solid-state additive manufacturing techniques offer various advantages when compared with other subtractive manufacturing processes in that it can manufacture neat-net shape of components of any shapes, complicated geometries and offered flexibility in customizations. The technique is new and have been successfully adopted by industries in the manufacturing of Al-, and Mg-alloy based components. The components fabricated with solid-state AM provide comparable mechanical and metallurgical properties with substructure manufacturing techniques. However, the technique is not yet standardized due to formations of several types of defects and poor surface finish. The common defects are crack, open channels, groove, internal cavities etc. which needs to be studied through a suitable technique. In the present chapter details of various destructive and non-destructive techniques have been discussed. Signal processing and image processing techniques can also be employed for the detection of defects in solid-state AM. Recent developments in technologies for defect identifications have also been discussed.

REFERENCES

[1] J.J.S. Dilip, G.D. Janaki Ram, Microstructure evolution in aluminum alloy AA 2014 during multi-layer friction deposition, *Mater. Charact.* 86 (2013) 146–151. https://doi.org/10.1016/j.matchar.2013.10.009

[2] R. Kawalkar, H. Kumar Dubey, S.P. Lokhande, Wire arc additive manufacturing: A brief review on advancements in addressing industrial challenges incurred with processing metallic alloys, *Mater. Today Proc.* 50 (2021) 1971–1978. https://doi.org/10.1016/j.matpr.2021.09.329

[3] R. Joey Griffiths, D.T. Petersen, D. Garcia, H.Z. Yu, Additive friction stir-enabled solid-state additive manufacturing for the repair of 7075 aluminum alloy, *Appl. Sci.* 9 (2019). https://doi.org/10.3390/app9173486

[4] Y. Du, D. Gu, L. Xi, D. Dai, T. Gao, J. Zhu, C. Ma, Laser additive manufacturing of bio-inspired lattice structure: Forming quality, microstructure and energy absorption behavior, *Mater. Sci. Eng. A.* 773 (2020) 138857. https://doi.org/10.1016/j.msea.2019.138857

[5] Z. Xie, Y. Dai, X. Ou, S. Ni, M. Song, Effects of selective laser melting build orientations on the microstructure and tensile performance of Ti–6Al–4V alloy, *Mater. Sci. Eng. A.* 776 (2020) 139001. https://doi.org/10.1016/j.msea.2020.139001

[6] R.S. Mishra, R.S. Haridas, P. Agrawal, Friction stir-based additive manufacturing. *Sci. Technol. Weld. Join.* 27 (2022) 141–165. https://doi.org/10.1080/13621718.2022.2027663

[7] R.S. Mishra, S. Palanivel, Building without melting: a short review of friction-based additive manufacturing techniques, *Int. J. Addit. Subtractive Mater. Manuf.* 1 (2017) 82. https://doi.org/10.1504/ijasmm.2017.10003956

[8] D. White, Object consolidation employing friction joining, *Google Pat..* 1 (1999). https://patents.google.com/patent/US6457629B1/en

[9] S. Palanivel, H. Sidhar, R.S. Mishra, Friction Stir Additive Manufacturing: Route to High Structural Performance, *JOM* 67 (2015) 616–621. https://doi.org/10.1007/s11837-014-1271-x

[10] R.S. Mishra, R.S. Haridas, P. Agrawal, Friction stir-based additive manufacturing, *Sci. Technol. Weld. Join.* 27 (2022) 141–165. https://doi.org/10.1080/13621718.2022.2027663

[11] A. Squillace, A. De Fenzo, G. Giorleo, F. Bellucci, A comparison between FSW and TIG welding techniques: modifications of microstructure and pitting corrosion resistance in AA 2024-T3 butt joints, *J. Mater. Process. Technol.* 152 (2004) 97–105. https://doi.org/10.1016/j.jmatprotec.2004.03.022

[12] S. Palanivel, P. Nelaturu, B. Glass, R.S. Mishra, Friction stir additive manufacturing for high structural performance through microstructural control in an Mg based WE43 alloy, *Mater. Des.* 65 (2015) 934–952. https://doi.org/10.1016/j.matdes.2014.09.082

[13] A.B. Tuncer, Nihan, solid-state metal additive manufacturing: a review, *JOM.* 72 (2020) 3090–3111. https://doi.org/10.1007/s11837-020-04260-y

[14] C. Parikh, R. Ranjan, A.R. Khan, R. Jain, R.P. Mahto, D. Chakravarty, S. Pal, S.K. Pal, Volumetric defect analysis in friction stir welding based on three dimensional reconstructed images, *J. Manuf. Process.* 29 (2017). https://doi.org/10.1016/j.jmapro.2017.07.006

[15] R. Ranjan, A.R. Khan, C. Parikh, R. Jain, R.P. Mahto, S. Pal, S.K. Pal, D. Chakravarty, Classification and identification of surface defects in friction stir welding: An image processing approach, *J. Manuf. Process.* 22 (2016) 237–253. https://doi.org/10.1016/j.jmapro.2016.03.009

[16] W.J. Arbegast, A flow-partitioned deformation zone model for defect formation during friction stir welding, *Scr. Mater.* 58 (2008) 372–376. https://doi.org/10.1016/j.scriptamat.2007.10.031

[17] M.C. Brennan, J.S. Keist, T.A. Palmer, Defects in metal additive manufacturing processes, *J. Mater. Eng. Perform.* 30 (2021) 4808–4818. https://doi.org/10.1007/s11665-021-05919-6

[18] R.R. Dehoff, S.S. Babu, Characterization of interfacial microstructures in 3003 aluminum alloy blocks fabricated by ultrasonic additive manufacturing, *Acta Mater.* 58 (2010) 4305–4315. https://doi.org/10.1016/j.actamat.2010.03.006

[19] X. Xu, X. Yang, G. Zhou, J. Tong, Microstructures and fatigue properties of friction stir lap welds in aluminum alloy AA6061-T6, *Mater. Des.* 35 (2012) 175–183. https://doi.org/10.1016/j.matdes.2011.09.064

[20] S.W. Park, T.J. Yoon, C.Y. Kang, Effects of the shoulder diameter and weld pitch on the tensile shear load in friction-stir welding of AA6111/AA5023 aluminum alloys, *J. Mater. Process. Technol.* (2017). https://doi.org/10.1016/j.jmatprotec.2016.11.007

[21] M.K. Yadava, R.S. Mishra, Y.L. Chen, B. Carlson, G.J. Grant, Study of friction stir joining of thin aluminium sheets in lap joint configuration, *Sci. Technol. Weld. Join.* 15 (2010) 70–75. https://doi.org/10.1179/136217109X12537145658733

[22] H. Liu, Y. Hu, C. Dou, D.P. Sekulic, Materials Characterization An effect of the rotation speed on microstructure and mechanical properties of the friction stir welded 2060-T8 Al-Li alloy, *Mater. Charact.* 123 (2017) 9–19. https://doi.org/10.1016/j.matchar.2016.11.011

[23] R.P. Mahto, C. Gupta, M. Kinjawadekar, A. Meena, S.K. Pal, Weldability of AA6061-T6 and AISI 304 by underwater friction stir welding, *J. Manuf. Process.* 38 (2019) 370–386. https://doi.org/10.1016/j.jmapro.2019.01.028

[24] Y. Chen, X. Peng, L. Kong, G. Dong, A. Remani, R. Leach, Defect inspection technologies for additive manufacturing, *Int. J. Extreme. Manuf.* 3 (2021) 022002, (2021).

[25] K.M.T. Jess M. Waller, Regor L. Saulsberry, Bradford H. Parker, Kenneth L. Hodges, Eric R. Burke, Summary of NDE of Additive Manufacturing Efforts in NASA, *AIP Conf. Proc.* 1650, 51 (2015). https://doi.org/10.1063/1.4914594

[26] M.S. Xavier, S. Yang, C. Comte, A.B. Hadiashar, N. Wilson, I. Cole, Nondestructive quantitative characterisation of material phases in metal

additive manufacturing using multi-energy synchrotron X-rays microtomography, *Int. J. Adv. Manuf. Technol.* 106 (2020) 1601–1615.

[27] S.K. Everton, M. Hirsch, P.I. Stavroulakis, R.K. Leach, A.T. Clare, Review of in-situ process monitoring and in-situ metrology for metal additive manufacturing, *Mater. Des.* 95 (2016) 431–445. https://doi.org/10.1016/j.matdes.2016.01.099

[28] J.L. Bartlett, F.M. Heim, Y. V. Murty, X. Li, In situ defect detection in selective laser melting via full- fi eld infrared thermography, *Addit. Manuf.* 24 (2018) 595–605. https://doi.org/10.1016/j.addma.2018.10.045

[29] S. Ekanayake, S. Gurram, R.H. Schmitt, Depth determination of defects in CFRP-structures using lock-in thermography, *Compos. Part B.* 147 (2018) 128–134. https://doi.org/10.1016/j.compositesb.2018.04.032

[30] Y. Ma, Z. Hu, Y. Tang, S. Ma, Y. Chu, X. Li, W. Luo, L. Guo, X. Zeng, Y. Lu, Laser opto-ultrasonic dual detection for simultaneous compositional, structural, and stress analyses for wire + arc additive manufacturing, *Addit. Manuf.* 31 (2020) 100956. https://doi.org/10.1016/j.addma.2019.100956

Chapter 8

Laser-based post-processing technologies for additive manufactured parts

Shrushti Maheshwari, Ashish Siddharth and Zafar Alam
Indian Institute of Technology (Indian School of Mines) Dhanbad,
Dhanbad, India

Faiz Iqbal
University of Lincoln, Lincoln, United Kingdom

Dilshad Ahmad Khan
National Institute of Technology Hamirpur, Hamirpur, India

8.1 INTRODUCTION

Additive manufacturing (AM) has characteristics such as low buy-to-flow ratio, a wide range of material fabrication, the ability to print intricate designs, and lower lead time. As a result, AM processes have been applied in a range of domains, including electronics, medical, aerospace, construction and even space technologies. It therefore becomes necessary for products manufactured by AM techniques to meet industrial requirements, including excellent surface finish, industrial functional tolerances, refined surface morphologies and adequate strength, density and toughness. Henceforth, the product manufactured using AM cannot be used directly and would require post-processing to eliminate the defects occurring during AM fabrication and to enhance surface characteristics. Several conventional methods have been utilized in the finishing of operations, principally mechanical, chemical, manual and electrochemical ones, as explained in Figure 8.1.

However, conventional finishing methods require skilled labour and high operational costs and are also time-consuming. Moreover, these processes could not be automated. Therefore, an unconventional technique known as laser polishing has been explored in recent decades. This study presents a detailed review of laser-assisted finishing on AM products. Efforts have been made to study the different techniques available, the crucial parameters and their effects on the finished product.

DOI: 10.1201/9781032616025-8

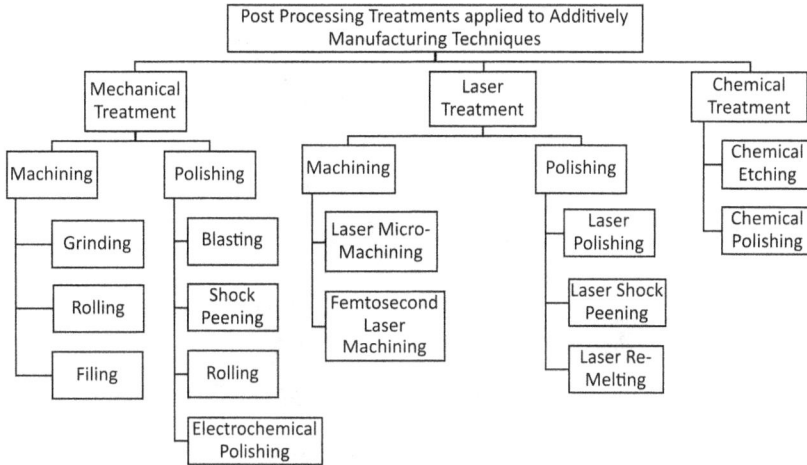

Figure 8.1 Post-processing techniques applied to additive manufactured products.

8.2 NEED FOR POST-PROCESSING FOR ADDITIVELY MANUFACTURED PARTS

Several issues need to be addressed before the additive-manufactured component can meet industrial functional tolerances, required surface characteristics, acceptable porosity, strength, and density. Additive manufacturing techniques, materials they are compatible with and the common defects are described in the following sub-sections.

8.2.1 PBF – Powder Bed Fusion

The Powder Bed Fusion process is one of the advanced additive technology processes for which research is currently underway. In this process, a concentrated heat source, either in the form of laser or thermal, acts on a sintered powder metal, fusing them together. This system requires the spreading of powder material over the previous layer once it is hardened, as shown in Figure 8.2. There are specific mechanisms for recoating, such as a roller or a blade, to provide fresh powder over the last layer.

The significant advantage of this process is that it does not require a support structure. This allows the complex part to be printed quickly, as post-processing is both less costly and relatively easy. However, some of the disadvantages of the process include hollow details, and that fine-channel post-processing will not be easy. It would be difficult to remove the trapped powder present inside such components. Also, the powders must be kept in an inert gas environment to prevent powders from getting oxidized, which increases production costs [1].

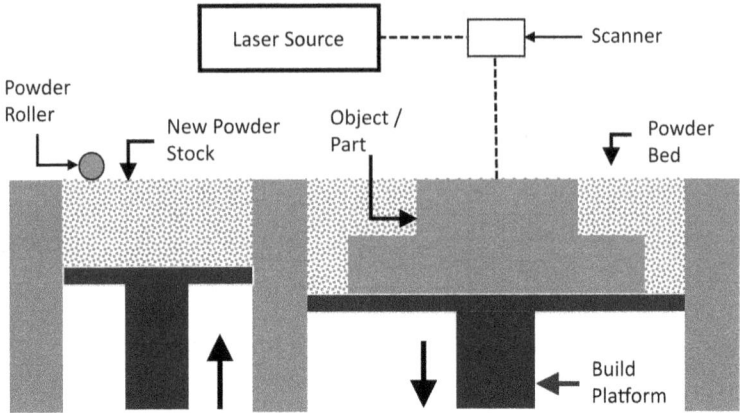

Figure 8.2 Schematic of the powder bed fusion manufacturing process.

The characteristics of powder for this process are:

1. Powdered particles should be smaller for better surface finish and accuracy.
2. The smaller-sized particle ensures a thinner layer.
3. The surface energy of particles should be higher.
4. The powdered particles should be kept in an inert environment to prevent contamination.
5. The powder particles should be light in weight.

8.2.1.1 SLM – Selective Laser Melting

Selective Laser Melting is a powder bed fusion process used to fuse industrial materials such as powdered plastics, polymers, rubbers, ceramics, composites or metals to create 3D printed parts. A laser is used as a heating source element, which melts and bonds polymer powder together. These powders are kept in an inert environment to prevent contamination, preferably nitrogen or argon. The powders must also be held at a specific temperature, lower than the material's melting temperature. This is usually done to avoid cussing or wrapping powdered elements and provide a moisture-free environment [2]. Direct Metal Laser Sintering is a powder bed fusion process and a particular case of selective laser melting. Here, only metal-based powder is used to create 3D-printed parts.

Based on the primary binding mechanism, SLM has been further divided into Chemically Induced Sintering, Solid-State Sintering, Full Melting and Liquid Phase Sintering. The powdered particle is fused at elevated temperatures without melting in the solid-state sintering mechanism. In this case, the fundamental driving force is the minimization of the total free energy of the particle. The major disadvantage of this process is that the powdered material is sensitive to temperature. In a Chemically Induced Sintering mechanism, thermally activated binders are used to fuse two particles. This binding machine is generally used in the case of ceramic parts. In this case, the only issue is high temperature and parts porosity, which is way higher than others. Liquid Phase Sintering is widely used in the glass industry. Both the structural material and the binder are fused at the melting point in the liquid phase of sintering. The mixture produces objects with a better surface finish, high density and uniformity. Lastly, Full Melting Sintering is used for Ti, Stainless Steel, Co-Cr alloy and some semi-crystalline materials such as nylon polyamide. In this case, the heat source's entire region of material impinged is melted to exceed the layer thickness. This method produces a high-density and well-bonded structure.

Need for post-processing of SLM manufactured parts: SLM manufactured parts require suitable post-processing techniques in order to eradicate various defects occurring in the product during manufacturing. These defects are discussed below.

Common defects occurring in Selective Laser Melting (SLM):

1. Defects due to powder used: The loose powder sometimes adheres to the surface during sintering and gets trapped inside. These loose powders create problems during post-processing. By contrast, in the heat-affected zone, the unused powder gets semi-bonded, and partial melting of these powders takes place due to conduction.
2. Defects resulting from improper scan speeds: High scan speeds and low liquid front rates produce low, dense parts, whereas low scan speeds in combination with high liquid fronts result in shrinkage and the crumbling of components.
3. Stair-stepping effect: The stair casing effect originates in a component with a built orientation greater than 45°. The main reason behind this is the splash occurring due to the high laser absorption of the laser used. This defect can be avoided by keeping a smaller layer thickness.
4. Balling melts: Few ellipsoidal and globular objects become detached from the melted zone; these objects are known as balling melts. Improper cooling of the melt pool and inappropriate scan speeds leads to balling melts.
5. Other defects include surface bumps, marked patterns, surface irregularities and liquid splash.

To eradicate these defects various post-processing techniques are utilized. Based on the mechanism used these techniques are described in Table 8.1.

Table 8.1 Need for post-processing in selective laser melting manufactured parts

Post processing mechanism	Post-processing technique	Need for post-processing
Thermal	Heat-Treatment, Solution heat-treatment, Hot iso-static Pressing	1. To reduce the residual stresses 2. To improve fatigue resistance
Mechanical	Abrasive Flow Machining, Friction Stir Processing, Shock Peening	1. To reduce the surface defects due to balling effects and powder addition 2. To increase surface integrity
Laser	Laser Polishing, Laser Shock Peening	1. To improve the mechanical and microstructural properties. 2. To induce favourable compressive residual stress

8.2.1.2 EBM – Electron Beam Melting

Electron Beam Melting uses a high-density electron beam to fuse the metal powder. These electron beams are emitted through a tungsten filament, which is focused and controlled with the help of magnetic coils, as shown in Figure 8.3. This method produces highly dense parts and is widely used for rapid prototyping and small-scale production in aerospace industries. Refractory and resistant materials such as tantalum, niobium, tungsten, vanadium, titanium and zirconium are processed using this method.

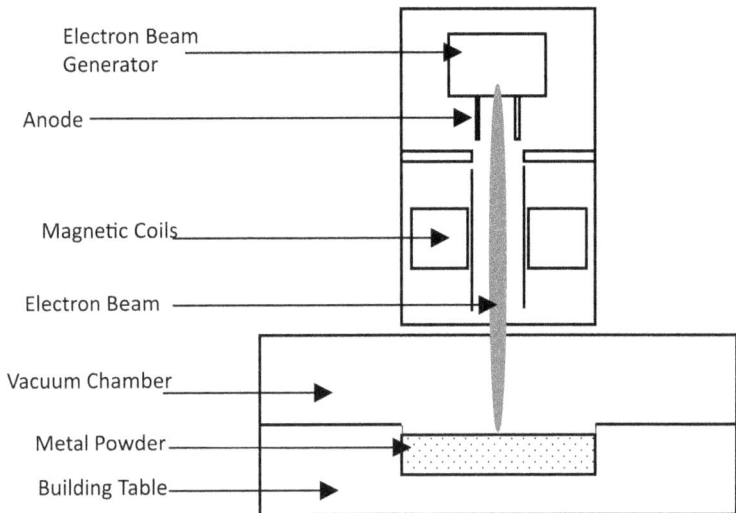

Figure 8.3 Schematic of Electron Beam Melting.

The system scans a thin layer of metal powder, building up structures in a bottom-up format. These metal powders are kept under a vacuum to prevent highly reactive materials from oxidizing. In contrast to the laser beam, the electron beam process is highly efficient and less costly. Other advantages include a higher buy-to-fly ratio, printing multi-piece assembly as a single component, and printing hollow parts. Nevertheless, this system also has a significant drawback; the metal powder beds in this process are needed to keep at higher temperatures [3].

Need for the post-processing of EBM-manufactured parts:

In general, parts produced using EBM are observed to be denser than SLM parts; however, they have a poor surface finish, less yield strength and ultimate tensile strength and have almost comparable hardness. Defects occurring in EBM are discussed below.

1. Process-induced defects: Process-induced defects include thermal expansion of the material, instabilities in the melt pool, warping, electrostatic repulsion, and vaporization of the metallic material.
2. Power handling defects: This includes defects occurring due to improper ratio of fresh powder to reusable sintered powder, inappropriate amount of powder supplied and the inaccurate ratio of surface area to volume of powder particles.

Various post-processing techniques are executed with the purpose of removing these defects. Shui et al. [4] applied heat treatment and hot isostatic pressing on Ti6Al4V samples produced using EMBed. It was observed that tensile strength, as well as fatigue limit, increased. Moreover, samples post-processed with heat treatment showed better tensile properties compared to HIPed Ti6Al4V, whereas the fatigue limit was high in the HIPed sample.

8.2.2 FDM – Fused Deposition Modelling

Fused Deposition Modelling is an extrusion-based low-cost 3D printing technology which is widely used for rapid prototyping. This method extrudes material through a moving nozzle in the x and y direction, depositing material layer by layer in the z-direction, as shown in Figure 8.4. The nozzle mechanism is usually programmed and controlled to allow material flow. The material is supplied in the form of a roll of plastic filament. The filament is then converted to a semi-liquid state by passing it through a temperature-controlled extrusion head-set [5]. The method uses engineering-grade plastics such as ABS, PLA or 9085 Resin to create 3D-printed structures.

Common defects occurring in Fused Deposition Modelling (FDM): The major setback of this technique is its limitation on the availability of printing materials. The printed materials are weak and have anisotropic properties, leading to the several defects discussed below. It is complicated to eliminate defects with pre-processing. Therefore, proper post-processing is required.

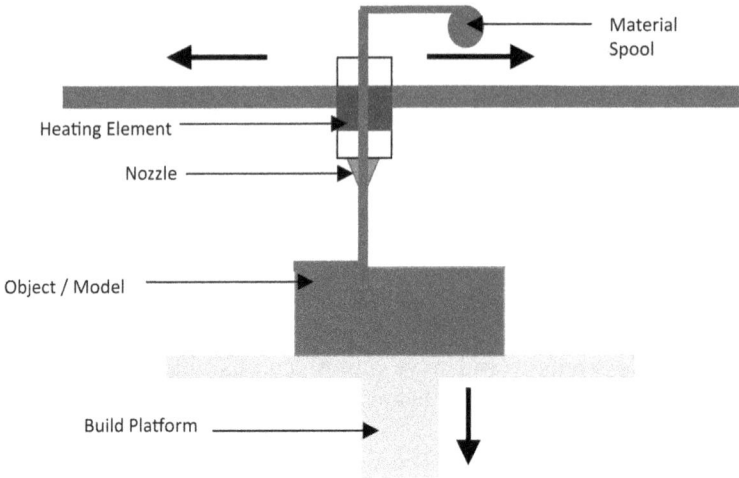

Figure 8.4 Schematic of fused deposition modelling.

1. Weak adhesion of the bottom layer: Due to the temperature gradient between the bed and the first layer to be extruded, adhesion is not as strong as between the layers. Thereby, warpage is observed.
2. Material delamination: This is the result of inconsistency between the feed rate and the rate of supplied extruded wire. Also, the two consecutive layers get deboned or de-laminated due to improper adhesion. In the case of composites, fibre pull-out takes place when the FDM-formed composites are loaded, as shown in Figure 8.5.
3. Chordal error: Chordal is the difference between the curvilinear path desired and the actual tessellation path shown in Figure 8.6.
4. Other defects: Debonding between two filaments, and stair-case defects are shown in Figure 8.7.

Post-processing techniques available for FDM parts: To eliminate the defects discussed above following post-processing techniques are described in Table 8.2.

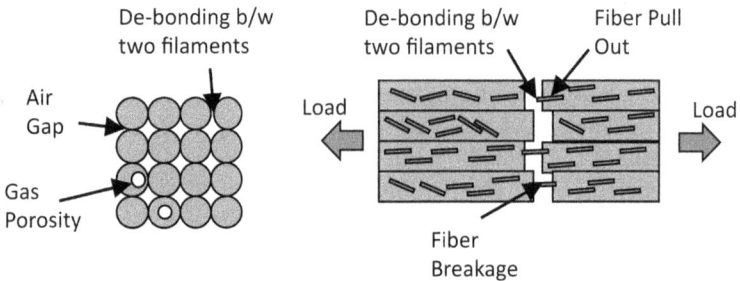

Figure 8.5 Debonding between two filaments and failure of composites.

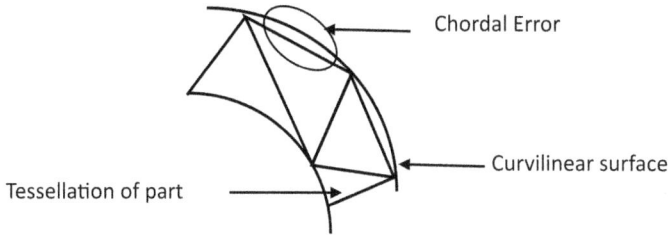

Figure 8.6 Chordal error.

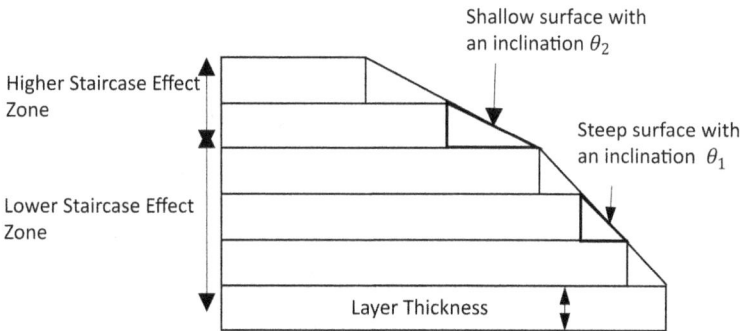

Figure 8.7 Staircase defect.

Table 8.2 Need for post-processing in fused deposition modelling manufactured parts

Post processing mechanism	Post-processing technique	Need for post-processing
Thermal	Heat-Treatment, Hot-Cutting Machining	1. To improve the tensile strength of FDM-manufactured parts, 2. Reducing the void present in part, 3. For improving surface roughness 4. To reduce the staircase effect
Mechanical	Abrasive Milling, Abrasive Flow Finishing, Polishing, Ball Burnishing, Sand Blasting	1. To Improve surface finish and continuity 2. To increases hardness 3. To relive stress concentration
Laser	Laser Polishing, Laser Shock Peening	1. Improves dimensional accuracy, slope and surface continuity 2. Prevails strength against delamination. 3. Induces compressive stress
Chemical	Vapour Smoothing, Epoxy Treatment	1. To improve surface roughness

8.2.3 Binder Jetting

Binder Jetting adds a new branch to additive printing techniques. The process is somewhat similar to SLM, with the only difference being that no heating element is used. The absence of heating elements makes its operation cost cheaper than other techniques. In this process, binder particles are sprayed through inkjet nozzles onto powdered materials, further bonding them to create a solid part. The substantial part is then left to be cured to gain strength. This process is advantageous when compared with other techniques as it can create low-cost 3D printed pieces using metal powder, sand and ceramics. Further, the bonding process of the powders occurs at room temperature. Hence distortions due to thermal effects are negligible [6].

8.2.4 WAAM – Wire Arc Additive Manufacturing

Wire Arc Additive Manufacturing (WAAM) is a type of Direct Energy Deposition technology, also called DED. It uses an electric arc as a heat source to melt the metal wire. This technology is also used to create 3D-printed metal components. Unlike PBF, which uses metal powder, the WAAM system is fitted in the end effector of a robotic arm. The arm gives the required shape by depositing the melted component on a substrate material, later cut off. This melted wire is extruded in the form of beads on the top of the substrate layer, as shown in Figure 8.8. These beads stick together to create a new layer of metal material. The process is repeated layer by layer until the desired component is manufactured.

Since the process uses metal wire, it can work with several materials such as stainless steel, titanium alloys, and aluminium alloys. The process is advantageous compared to PBF technology, as it can be used to manufacture more significant metal parts. PBF machines have a limited working envelope,

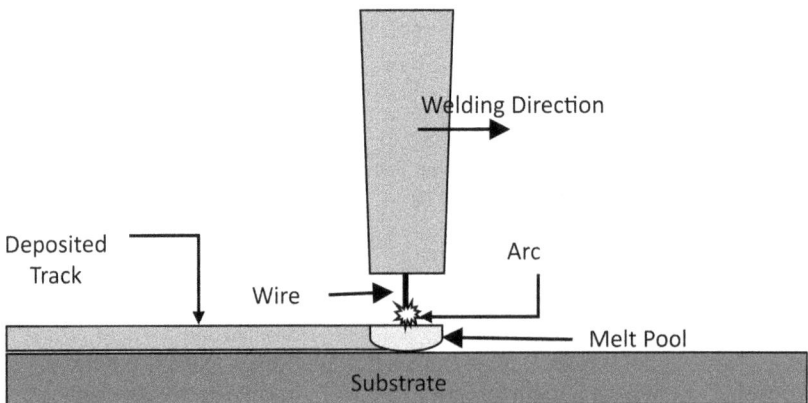

Figure 8.8 Schematic of wire arc additive manufactured process.

while WAAM can work inside a robotic arm workspace. This technology is also the least expensive as it does not require metal powder and has the least operation cost. The parts produced are of high density and have strong mechanical properties. The only challenge in this process face is the development of residual stress and distortions, due to which the manufactured component has a poor surface finish and requires post-processing.

Therefore, the need arises for post-processing techniques to enhance the properties, surface morphologies and texture of AM components. These can be achieved by employing suitable post-processing techniques and thermal or non-thermal treatments such as heat treatment, sandblasting, micromachining, chemical etching, and electro-polishing. Above mentioned conventional techniques are time-consuming and have limited applicability when it comes to intricate geometries and free-form surfaces. In recent decades, a latent substitute exploiting laser has been considered laser polishing [7].

8.2.5 Need for post-processing of parts formed using WAAM

WAAM is prone to certain defects which need to be addressed to obtain a finished part with desired material characteristics. Common defects occurring in WAAM and their causes are discussed below.

Due to variation in thermal energy, inappropriate process parameters and asperities present on surface various surface defects such as bulges, dents, pores, cracks, collapses are prevalent in products manufactured by AM. Weld profile variability is also observed in parts produced using WAAM. Due to the thermal gradient existing during the process, unfavourable residual stresses get accumulated leading to varying grain sizes.

In order to remove these defects and further enhance surface as well as material characteristics, various post-processing techniques are utilized. In a study, WAAMed Al 2219 alloy (Al-Cu alloy) was subjected to heat treatment and it resulted in an improvement in yield strength and ultimate strength [8]. In another post-processing study, WAAMed Al 2319 alloy was exposed to vertical and side rolling and it was observed that there was a significant reduction in residual stresses and deformation. Further, a considerable increase in hardness was also observed [9].

8.3 LASER-BASED FINISHING TECHNIQUES

Laser-based finishing is a non-contact surface-finishing operation that utilizes laser beams with suitable intensity to remove irregularities in the surface of objects under finishing. (Laser stands for light amplification by stimulated emission of radiation.) Thus, laser finishing functions by emitting highly intensified laser beams that contain sufficient energy to melt the material's surface under consideration. Laser-based finishing provides

```
                          ┌─────────────────────────┐
                          │  Laser Based Finishing  │
                          └─────────────────────────┘
          ┌─────────────────────┐          ┌─────────────────────┐
          │   Laser Polishing   │          │ Laser Shock Peening │
          └─────────────────────┘          └─────────────────────┘
  ┌──────────────────────┐  ┌──────────────────────┐
  │ Based upon Irradiation│  │ Based upon Surface Irradiated│
  │ Imparted by the Laser │  │    by the Laser Beam   │
  │         Beam          │  │                        │
  └──────────────────────┘  └──────────────────────┘
         ┌──────────────────┐      ┌──────────────────┐
         │ Continuous Wave  │      │   Surface-over   │
         │ Laser Polishing  │      │     Melting      │
         └──────────────────┘      └──────────────────┘
         ┌──────────────────┐      ┌──────────────────┐
         │ Pulsed Wave Laser│      │ Shallow Surface  │
         │    Polishing     │      │     Melting      │
         └──────────────────┘      └──────────────────┘
         ┌──────────────────┐      ┌──────────────────────┐
         │ Ultrashort Pulsed│      │ Laser Selective Localized│
         │      Wave        │      │ Ablation of rough peaks │
         └──────────────────┘      └──────────────────────┘
```

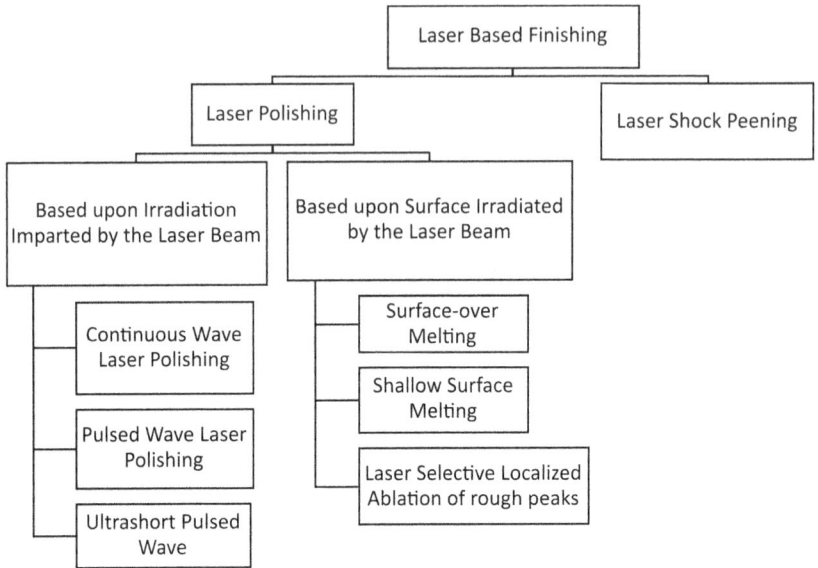

Figure 8.9 Classification of the laser-based finishing process.

high flexibility in terms of the range of systems, parameters variation, and materials to which it can be applied (stainless steel, material steel, titanium alloys, nickel alloys, cobalt chrome, diamond, silica, glass and some aluminium alloys). The process can be automated entirely, thereby reducing human effort and making it less time-consuming. Selective processing can also be accomplished using laser-based finishing. Different laser parameters, such as laser power, wavelength, spot size, scan speed repetition rate, and laser fluence, can be controlled to obtain versatility in laser-based finishing, remelting, and ablation of a wide range of materials. Classification of the laser-based finishing process is shown in Figure 8.9.

Laser-assisted finishing is done mainly by the following mechanisms:

8.3.1 Laser polishing

Laser polishing is a laser-based finishing operation which leverages the thermal energy generated by the laser to remove irregularities in the surface and enhance the surface morphologies of the workpiece.

8.3.1.1 Mechanism of laser polishing

During laser polishing, as the laser beam of adequate energy density impinges, the surface melting of microscopic layers takes place. A melt pool is developed, and the melted material inclines to restructure at the same horizontal level owing to the action of gravity and surface tension. As the laser beam moves away, temperature drops and the melted layer rapidly

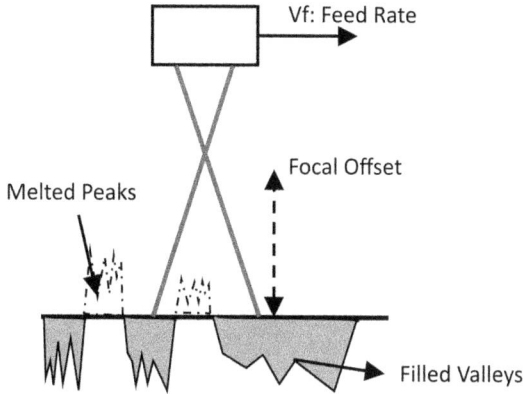

Figure 8.10 Melting of peaks and filling of valleys in laser polishing.

solidifies again. To reduce surface roughness, the depth of the affected layer must be sufficient to melt the peaks of the abnormalities. However, it should not be greater in depth than the valleys of the irregularities to be removed, as shown in Figure 8.10. Therefore, a controlled predetermined amount of energy should be imparted to the surface by the laser beam.

Microstructure, thermal properties, fluid dynamics and surface morphology of the polished laser components profoundly depend on the energy input during the laser polishing. Energy input can be varied by changing the laser beam's scanning speed and input power.

Based upon the irradiation imparted by the laser beam, laser polishing can be categorized into:

8.3.1.2 Continuous-wave laser polishing

Continuous-wave laser polishing uses layers that emit continuous beams rather than discrete beams. Since unremitting beams encroach the surface, a larger heat-affected zone is produced in this compared to pulsed wave laser polishing, as shown in Figure 8.11. These have the capability to invade the surface at a higher processing speed. Usually, continuous beam laser polishing finds its applicability in macro production. Nevertheless, the depth of the melted layer and heat-affected zone in this process reaches up to 100 microns [10, 11]. Thus, when it comes to micro-polishing, this may not have an advantage over pulsed wave laser polishing. In the case of continuous-wave laser polishing, the melt pool generated is affected by the scanning velocity of the laser, laser diameter, and the energy imparted to the workpiece, which depends on the laser power characteristics. Continuous-wave laser polishing finds its application in polishing a large number of materials. Wang et al. [12] employed a continuous-wave laser to laser clean fused quartz machined surfaces using a 25W CO_2 laser.

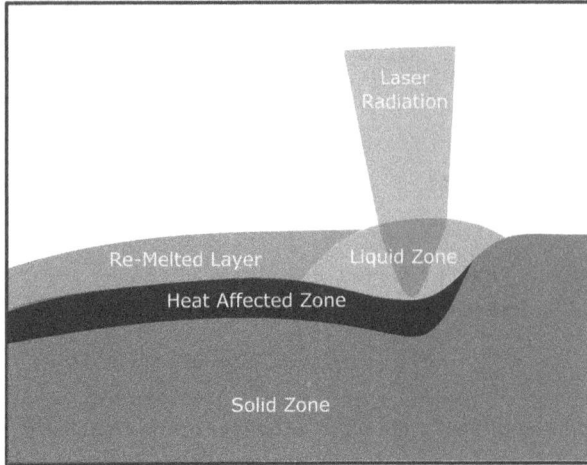

Figure 8.11 Schematic of continuous-wave laser polishing.

8.3.1.3 Pulsed wave laser polishing

Pulsed laser beam polishing employs discrete beams that intermittently impinge the surface under consideration. Once the beam hits the surface, melting takes place and thus melt pool is created. Forces acting due to surface tension and viscosity provide damped oscillation on this melted surface. Now, suppose the time taken by the oscillation to damp out is smaller than the time during which the surface remains molten. In that case, a smoother surface will be achieved upon solidification as described in Figure 8.12 [13]. Then, the solidification of the melted layer takes place before the next beam impounds the surface. Thereby, the laser beam's intensity, pulse duration, and percentage overlap of the laser become one of the most significant

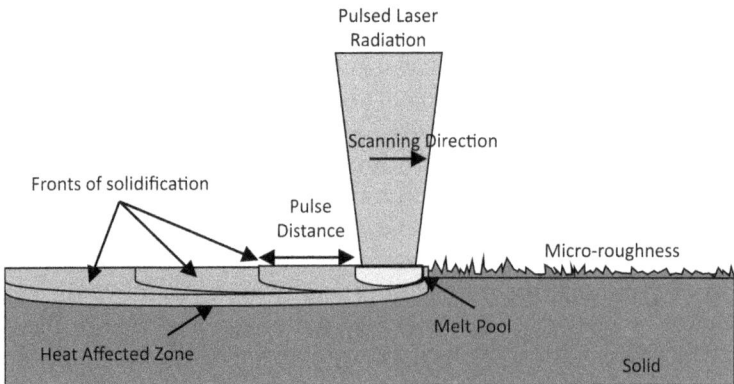

Figure 8.12 Schematic of pulsed wave laser polishing process.

process parameters. Since discrete beams are used, a significant reduction in laser power serves as an upper hand compared to continuous wave pulse polishing. Pulsed laser beams are primarily used in cases where micro-polishing is required.

Pulsed wave laser-based polishing investigated with a spot size of a few hundred microns and a pulse duration of 200 ns has yielded positive outcomes. Perry et al. [14] utilized pulsed layer micro polishing with a spot size of 60 μm along with a pulse duration of 300–650 ns and effectively reduced the surface roughness of micro-milled and microfabricated parts. It has been demonstrated by Willenborg et al. [15] that short-wavelength (high spatial frequency) features are directly influenced by shorter laser pulse durations. Perry et al. [16] came up with an analytical model that predicts the various laser durations that will be required to smoothen different surface features. They also projected that a melt depth of approximately 200 nm can be achieved using laser pulses of a duration of 650 ns.

The surface finish in the pulsed laser micro polishing can be improved by implementing thermocapillary flow [17]. By prolonged pulse duration, thermocapillary flows can be generated due to thermal gradient-inducing surface tension effects. These flows significantly reduce the surface roughness, but at the cost of producing spatial frequency surface features. Nevertheless, these features could be removed by repolishing the surface with short pulse durations in the absence of thermocapillary flows. Culminating these two processes, a reduction in surface roughness can be obtained without even producing spatial frequency features. When implemented on Ti-6Al-4V, this process was capable of reducing the surface roughness by 72% of the original roughness.

8.3.1.4 Ultra-short pulsed or femto-second based laser polishing

Continuous-wave, micro pulse-based as well as nano pulse-based laser necessitates high-control laser wavelength/material in order to ensure that the melting layer gets an adequate amount of energy absorption to accomplish melting up to the required depth. The amount of thermal energy facilitated by using continuous-wave, micro- and nano-pulsed laser also results in unfavourable structure changes together with high heat affected zone, larger melt depth thickness, melt front solidification leading to the formation of high spatial frequency ripples, and errors arising due to flow of molten material [18–20]. Thus, a non-contact, high-accurate laser-based polishing using femtosecond lasers was introduced. Femtosecond lasers have the ability to impart pulses with a time duration of 10^{-15} seconds. In so doing, a precise, localized ablation-based material removal with a nominal thermal impression on the surface can be achieved [21, 22]. Taylor et al. [23] investigated freeform model-guided femtosecond laser polishing of germanium and claimed to attain single-digit nano-meter optical surface quality while

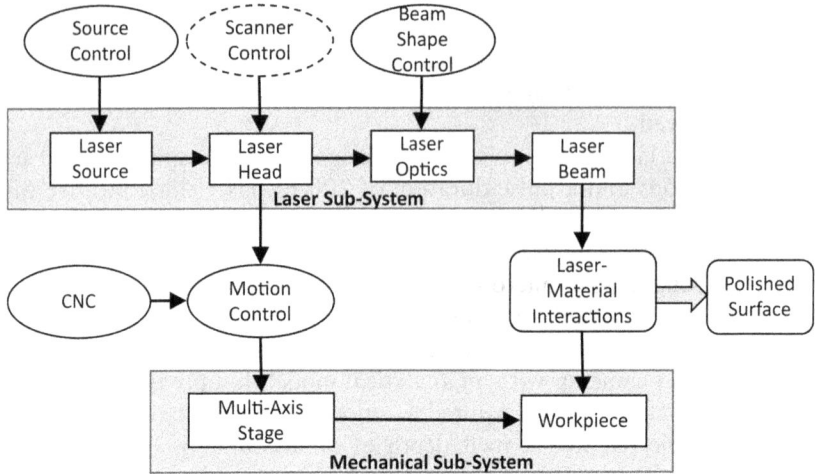

Figure 8.13 Working mechanism of laser polishing.

maintaining high precision of material removal of the contour under polishing. Ruck et al. [24] conducted a femtosecond laser-based single-track ablation test on cubes of AlSi10Mg0.4. These cubes were fabricated using a laser powder-based fusion technique. It was observed that scan velocity and repetition rates of pulse strongly influence the ablation depth achieved. Figure 8.13 presents the working mechanism of the laser polishing process.

Based on the amount of laser energy imparted, the region of surface it is imparted to, and the thickness of the melt layer concerning peak-to-valley distance of roughness laser polishing mechanism can be categorized as a surface over melting (SOM); laser selective localized ablation of rugged peaks and shallow surface melting (SSM).

8.3.1.5 Surface over melting

In surface over melting (SOM), a large amount of laser is imparted to the surface, which leads to the melting of not only the cap but also the whole particle [25]. Subsequently, the melt pool is created, leading to the formation of ripples shape on the material's surface. The excessive amount of surface melting induces a surface tension gradient resulting in the formation of capillary waves along with surface ripples, as shown in Figure 8.14.

8.3.1.6 Shallow surface melting

In shallow surface melting (SSM), a smaller amount of laser energy is used when compared with SOM, and the melt layer thickness is kept smaller than the peak-to-valley thickness to ensure that the peaks of the surface roughness go under complete melting and the melt material from the peak fills up

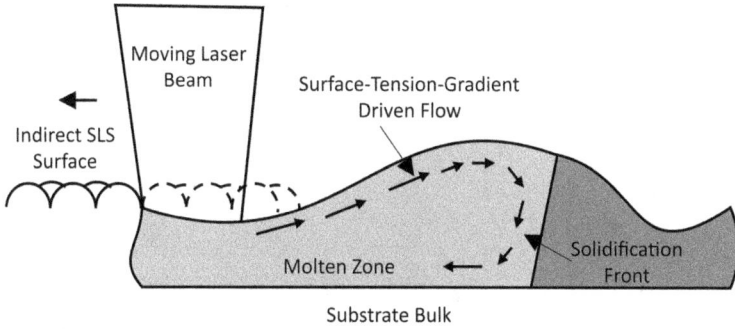

Figure 8.14 Schematic of surface over melting process.

the valley. During melting, the local curvature of peaks becomes prominent then the effective spreading of melted particles takes place towards the valley due to the tendency of liquid to minimize the local curvature. This driving mechanism is caused by capillary pressure created due to a reduction in viscosity. If the laser beam strikes and moves faster, then the depth of melting achieved is low and this safeguard smoothing the surface to a great extent [26].

8.3.1.7 Laser selective localized ablation of rough peaks

Selective laser polishing can be considered a subset of laser polishing. A locally predetermined region is polished by imparting continuous laser waves or pulsed laser waves while keeping the other regions unaffected, as shown in Figure 8.15. A dual-gloss effect is obtained due to a visual imprint created when a selective laser-polished surface is juxtaposed against the unpolished surface. A reduction in surface roughing of the lateral wavelength of about

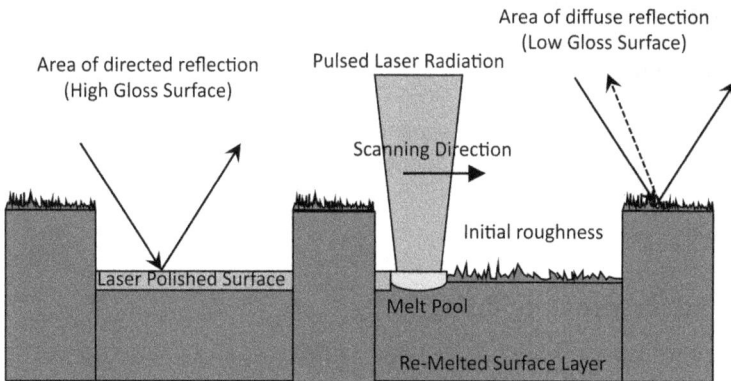

Figure 8.15 Schematic of selective laser melting.

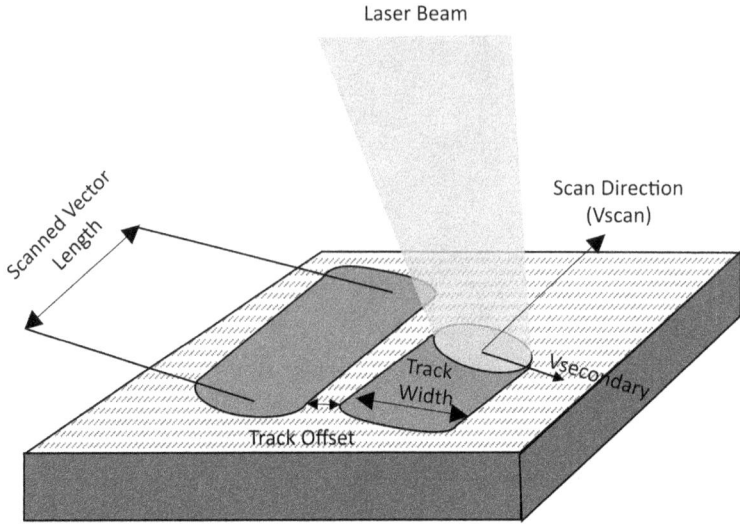

Figure 8.16 Laser scanning schematic.

80 microns can be achieved at the particular polished region, and hence a gloss is evident [27].

Velocity in the direction of the scan and the secondary direction, as shown in Figure 8.16, is related by Equation 8.1 given below.

$$V_{sec} = dV \times \frac{V_{scan}}{(x + dy)} \tag{8.1}$$

where,

V_{scan} = Velocity in the primary direction and,
V_{scan} = Velocity in the secondary direction.
dy = Track offset
x = Scanning length

Laser power is modulated such that in some places, only travel takes place, whereas, in others, enough energy is provided for the surface to remelt and provide laser polishing. And thus, selective finishing takes place, and a dual gloss effect is observed.

8.3.2 Laser shock peening

One of the defects while using laser additive manufacturing is the occurrence of tensile residual stress in the component produced. This is inherently due to the temperature gradient owing to rapid heating and cooling during the process.

Therefore, to eradicate the defect laser shock peening (LSP), an advanced laser surface treatment is used, which conveys advantageous compressive residual stresses to the component. Consequentially, increasing the product life by providing resistance to fatigue and wear and other surface-related failures.

8.3.2.1 Mechanism of laser shock peening

The component to be laser shock peened is coated with a pain, black tape or other sacrificial material, as shown in Figure 8.17. This layer is exposed to highly intense and short laser pulses in the presence of a transparent water

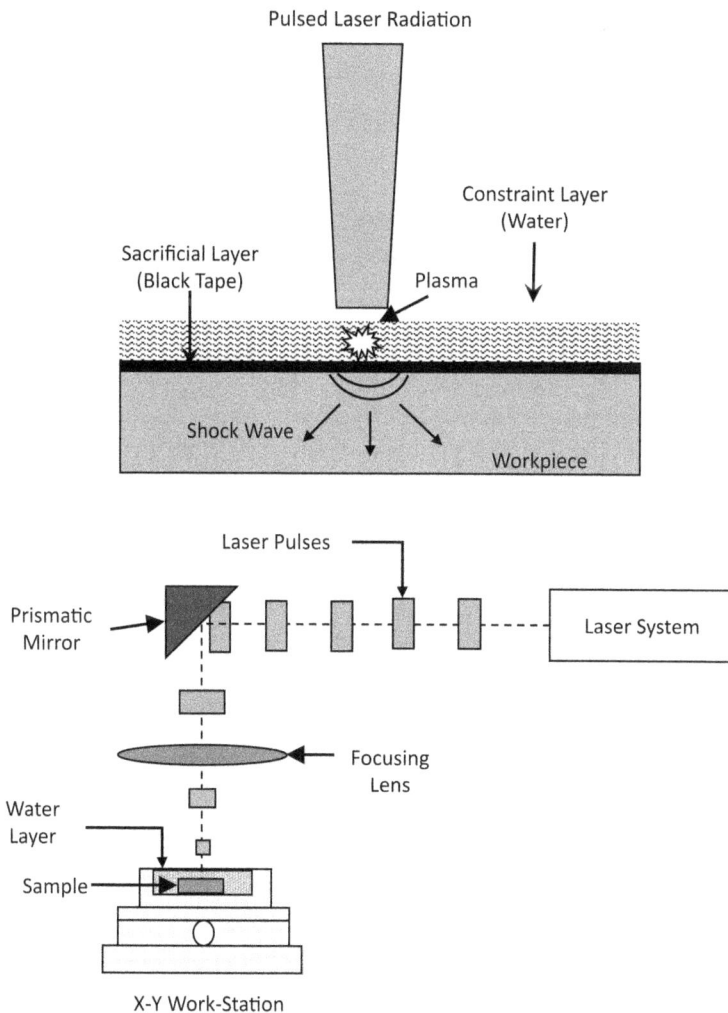

Figure 8.17 Schematic of laser shock peening mechanism.

Table 8.3 Laser shock peening parameters and their effect on the workpiece

Laser parameters	Workpiece parameters	Process parameters	Properties of the affected workpiece
Energy density of the laser	Materials of workpiece	Angle of impingement of laser beam	Inducement of compressive residual stress
Duration of laser pulse	Shape and curvature of workpiece	Laser handling system	Grain refinement
Spot size of the laser beam		Control system	Surface finish
Pulse rate of the laser beam		Scanning speed	Density enhancement
		Scanning direction	Stronger interfacial adhesion between layers
		Spot pattern	

layer. When high-intensity laser pulses impound on the sacrificial layer, it absorbs the energy and vaporizes immediately, and the remaining energy is converted into plasma. Expansion and sudden bursting of plasma plumes lead to the formation of a shock wave that proliferates into the workpiece. The pressure of the shock waves escalates, and when it exceeds the Hugoniot elastic limit of the material, yielding followed by plastic deformation occurs [28]. The plastic deformation inculcates compressive residual stress in the material's surface, sequentially enhancing the surface properties

Important laser shock peening parameters and the effect of laser shock peening on the workpiece treated can be described as shown in Table 8.3.

8.4 LASER POLISHING TREATMENT ON ADDITIVE MANUFACTURED COMPONENTS

Owing to the capabilities of laser polishing, it has been implemented in the post-processing of various components produced using AM. The following section describes laser polishing treatment on additive manufactured components.

8.4.1 Al-PLA composite components

(a) Fused Deposition Modelling
Due to extrusion-based layer-by-layer deposition in fused deposition modelling (FDM), various defects need to be addressed as described in Section 8.2.2 before the product can meet the end requirements.

Chen et al. [29] employed laser polishing with continuous-wave to augment the surface morphology and the quality of aluminium fibre-polylactide acid (Al-PLA) composite parts fabricated using FDM. The mechanism involved was when the laser beam hits the composite specimen surface, its anisotropic surface starts to melt, the polymer chains break down, and the melt pool is developed. The molten material from the crest of the roughness flows towards the valley due to gravity and surface tension, resulting in uniformity on the surface. After the passing of the laser beam, the surface affected cools down, and the melt pool solidifies, resulting in re-bonding of the material, and a finished denser specimen was obtained. It was stated that the roughness reduced to 0.32 μm from its initial value of 5.64 μm, and a stronger interfacial adhesion was found between the PLA matrix and Al fibres.

8.4.2 Aluminium components

(a) Laser-Powder based Fabrication

Hofele et al. [30] studied the mechanical and surface morphological properties after laser polishing aluminium AlSi10Mg manufactured using laser powder-based fabrication. A tensile test was performed on the laser-polished post-process part to investigate changes in the mechanical property as well as microstructural changes in the melted zone. It was observed that larger melt pools accompanied by slow cooling rates lead to a coarser eutectic structure when assessed against the initial untreated specimen. The lamellar grains are transformed into uniform grains. Yield strength and tensile strength increased after laser post-processing and reached their original value at the initiation of L-PBF.

8.4.3 Steel alloy components

(a) Selective Laser Melting (SLM): Solheid et al. [31] explored the variation of two crucial process parameters, viz., the scan velocity and the repetition rate on the SLM manufactured 18 maraging 300 steel. An ultrafast laser fibre system was used for imparting femtosecond laser radiations. It was observed that upon decreasing the scanning speeds, surface roughness tends to increase, whereas when the scanning speed was increased, the surface roughness was found to be similar to that of the inbuilt AM component. However, nano undulations were also acknowledged as laser-induced periodical surface structures (LIPSS) formed in the case of high scanning velocities. Thus, it was concluded that, owing to cold ablation, it is advantageous to employ femtosecond laser-based polishing for selective particle ablation and edge processing.

Hofele et al. [32] executed laser cleaning as well as laser polishing on an X2CrNiMo17-12-2 steel specimen which was fabricated using SLM. A short pulse laser was used for a laser-cleaning operation with a power rating of 20 W, whereas for laser polishing, a disk-type laser with a maximum power of 4000 W was used. Continuous-wave, as well as pulsed-wave polishing, were analyzed. A frequency of 1000 Hz with a pulse duration of 0.3 ns resulted in surface finish improvement from 12.5 to 0.35 μm.

(b) Binder Jet AM: Bhaduri et al. [33] analyzed the critical parameters of laser polishing on stainless steel (SS316L) manufactured by implementing binder jet AM on a commercially available DM (Direct metal) system. The initial roughness of the specimen tested, along with the energy density imparted during polishing and the pulse overlap laterally to the beam scanning direction, came out to be the most persuasive parameters during laser-based post-polishing. It was observed that if the energy density is kept too high, then ablation occurs, whereas if it is kept too low, improper melting resulting in an ununiform surface is witnessed. At the same time, optimized energy density resulted in reduction of surface roughness by 94% compared to its initial value. The higher the initial roughness higher is the potential for LP. The region with high laser density impingement; witnesses amalgamation of ablation as well as a surface over melting, resulting in higher porosity in the sub-melted layer. On the contrary, a dense sub-surface area is obtained if the energy impinged is less. Furthermore, porosity is eliminated.

8.4.4 Titanium alloy components

(a) Selective Laser Melting
Due to the vulnerability of Ti-6AL-4V to rapid oxidation and stress cracking [34], it seems complicated to polish. Ma et al. [35] performed laser polishing on Ti-based alloy Ti-6.5Al-3.5Mo-1.5Zr-0.3Si (TC11), and Ti-6Al-4V (TC6) manufactured using AM. White-Light interference, Focus ion beam, Scanning Electron Microscopy, Confocal Microscope, Focus Ion Beam, X-ray diffraction, and Energy Dispersive Spectrometer were used to investigate surface morphology as well as the subsurface cross-section before and after polishing. Results showed that initial surface roughness of more than 5 μm was reduced to less than 1 μm. α+β dual-phase structure was transformed into α'martensitic phase in both TC4 and TC11 due to rapid melting and solidification during LP. There was an increase in microhardness of both samples, of about 32% in TC4 and 42% in TC11. Marimuthu et al. [36] developed a computational fluid dynamic-based numerical model which helps to investigate pool dynamics. Energy density supplied came out as the most influencing parameter affecting the melt pool dynamics and controlling surface roughness. Excessive energy density led to carbonization and oxidation. When the melt pool

velocity is kept to a minimum, then it results in a good surface finish on a relatively larger track width. Melt pool velocity can be varied using laser speed and power. Continuous laser beams were used, and a reduction of roughness from 10.2 to 2.4 μm took place when the optimized parameters were used. To achieve a good surface finish, melt pool velocity should be kept to a minimum. Gora et al. [37] conducted a continuous wave laser of wavelength 1064 nm with a maximum power of 40 W was used for polishing on Ti-6Al-4V and studied the influence of laser parameters, initial surface conditions and scanning approaches on the obtained surface finish. Three different scanning's were tested for 1,2 and 3 cycles, where each cycle contains four polishing passes. The scanning's were 1. Standard raster scanning 2. Perpendicular scanning directions 3. Halftone printing angles. The best reduction was obtained in the case with the halftone scanning pattern and was about 21.5 times reduction of surface roughness for flat samples and 10 times for cylindrical samples.

8.4.5 Cobalt-Chromium components

(a) Selective Laser Melting

Cobalt-chromium alloys find their applicability in dentistry. They have been extensively used for implants, replacing bones, jaw plates, teeth, and other dental tissues. Nevertheless, it becomes essential to control the release of metal ions from these implants to inhibit toxicity in the mouth cavity environment. Therefore, CoCr used as dental implants must be corrosion-resistant, and it has been found that a stable oxide film coating provides corrosion resistance. [38]

Wang et al. [39] analyzed corrosion resistance, microstructure and surface morphology of laser-polished CoCr alloy, which was initially fabricated using selective laser melting. It was detected that laser polishing results in a substantial increase in the corrosive resistance of CoCr. The corrosion resistance was found to be reliant on the outer layer of oxide and inner structure. Laser power and distance of the workpiece from the laser were distinguished as the most influencing parameters in determining corrosion resistance and surface morphology. High laser powers and low distance up to a specific limit between the workpiece and laser result in the oxide layer's formation, thereby inculcating high corrosive resistance. It was also noted that as the laser power increased from 40% to 100%, surface roughness decreased from 4.976 to 0.45 μm, whereas upon increasing the object distance from 204 to 208 mm, surface roughness increased from 0.45 to 15.91 μm.

Gora et al. [40] also performed LP on SLM-manufactured Cobalt-Chromium. Initial conditions, scanning patterns and laser parameters were considered during the LP mechanism. In this case, four different scannings proceeded, viz. 1. Raster cycle, 2. Scanning with increasing angles, 3. Perpendicular scanning directions and 4. Halftone print angles. It was concluded that the best results were obtained in the case of Halftone printing angles, and an overall decrease of between 85% and 96% was observed based on the initial roughness.

8.5 LASER SHOCK PEENING TREATMENT ON ADDITIVE MANUFACTURED COMPONENTS

Capabilities of laser shock peening treatment have been exploited to postprocess different additive manufactured components as described in the following sections.

8.5.1 Inconel 718 components

(a) Direct Metal Laser Sintering (DMLS)
Jinoop et al. [41] performed and investigated laser shock peening on direct metal laser sintering (DMLS) built Inconel 718. Two essential parameters, viz. laser power and the number of shocks, were also optimized. It was observed that LSP changed the mechanical properties as well as the surface morphology of the part. They reported compressive residual stress after the laser shock peening was found to be 214.9–307.9 MPa and the wear improved by a factor of 1.7 compared to that of an initial untreated part. The study concludes that laser shock peening comes out as a potential method to improve surface texture and morphologies of complicated geometrical AM build parts.

8.5.2 Titanium alloys components

(a) Laser Additive Manufacturing (LAM): Luo et al. [42] applied laser shock peening on an alloy of titanium (TC17) which was manufactured using LAM. They measured the fatigue strength of the substrate specimen of TC17 and then the component after LAM. Again, fatigue strength was measured after laser spot peening post-processing. It was found that fatigue strength decreased to 365 MPa from its initial value of 405 MPa upon building it by LAM due to the occurrence of residual tensile stress and the formation of coarser grains in the heat-affected zone. However, after laser shock peening post-processing, it increased to 451 MPa. This is due to the elimination of tensile residual stress as well as a change in microstructure owing to the generation of the plastic zone, refined grains and increased dislocation density.

(b) Selective Laser Melting (SLM): Jiang et al. [43] analyzed the effects of the LSP on Ti6AL4V specimens produced by selective laser melting (SLM). It was concluded that LSP resulted in refined microstructure removal of tensile residual stresses. Also, in the ultra-high cyclic fatigue (UHCF) loading region, a lower S-N curve was obtained compared to specimens without undergoing LSP.

(c) Electron Beam Melting (EBM): Jin et al. [44] performed LSP on Ti-6Al-4V fabricated using Electron Beam Melting. During LSP, Nd: YAG laser emitting lasers of wavelength 1064 nm was used with the pulse duration kept as 2 ns and the diameter of the laser beam as 2.5 mm. It was concluded that upon performing LSP microhardness of the specimen increased by approximately 11%. Also, LSP intensely refines the microstructure of Ti-6Al-4V by forming α-phase sub-grains owing to the deformation twinning and dislocation phenomenon. The authors also established that the untreated EBM produced Ti-6Al-4V run-out fatigue strength for 20,00,000 fatigue cycles came to be 600 MPa, whereas it was increased to 700 MPa for the LSP post-processed Ti-6Al-4V for the same number of fatigue cycles. This was due to suppression of crack initiation and propagation, owing to α phase grain refinement.

(d) Wire and Arc Additive Manufactured (WAAM): The metal components manufactured using WAAM contain unfavourable residual tensile stress and large columnar grains due to thermal gradient and rapid solidification taking place during the fabrication process. Chia et al. [45] executed LSP on WAAM Ti-17 titanium alloy to eliminate these issues. That resulted in the generation of mechanical twins in coarse α phases, which eventually led to more refined α grains. A severe zone of plastic deformation was also observed, resulting in compressive residual stress measuring to a maximum of 740 MPa, which in turn increased the fatigue resistance. It was stated that microhardness increased approximately by 50 HV after the LSP post-processing; meanwhile tensile residual process after Wire and Arc reached almost zero.

8.5.3 Steel components

(a) Selective Laser Melting (SLM)
Kalenticsa et al. [46] employed LSP on two different grades of stainless steel prepared by SLM: 1. An austenitic 316L 2. Martensitic 15-5 precipitation-hardenable PH1. Various LSP parameters, viz. laser spot size, laser energy, and the overlapping of shocks, were changed to study their effect on the post-processed microstructure, surface morphology and fatigue strength. Treatments were performed with and without an ablative medium. The laser source used was a 7.2 ns Nd: YAG laser with a wavelength of 532 nm. It was concluded that laser peening proved to effectively change the residual stress profile by inducing favourable compressive stress. In the case of 316 L, LSP was able to change residual stress to compressive, and in the case of PH1 compressive residual stress increased and penetrated to a more significant depth. When used with ablative coating, smaller spot sizes led to greater depth and amount of residual compressive stress. Further, it was proposed that using LSP as an integration to the SLM in the

building phase itself may give additional depth and volume of compressive residual stress, and it is potential for selective treatment of specific areas can be leveraged.

8.6 SUMMARY AND PERCEPTION

Based on the chapter reviewed, the following comparison table of roughness data before and after the laser post-processing technique (Table 8.4) has been made for different materials and the technique used for laser polishing.

This data is visualized in the form of Figure 8.18. It can be observed that the higher the magnitude of the slope more remarkable the change in the surface roughness. Again, upon moving in the +y direction or upwards, the slope of the lines increases. That implies that as the initial roughness increases, the potential for change in roughness is tremendous. Additive components made by Ti alloys can be laser-finished to a significantly low roughness value compared to other materials.

This study provides a detailed review of laser-based post-processing techniques for additive-manufactured products. The following insights have been drawn based on the papers reviewed.

- Laser polishing techniques depend on the laser parameters (type of laser beams used; the beam's spot size, laser power, diameter, energy density, the distance between the laser beam and the object) and scanning parameters (scanning velocity, directions, and strategies). If these parameters are optimized based on the material to be polished, then a high finish, richly dense and surface with enhanced properties can be achieved.
- Surface finishing is found to be dependent on the initial roughness value.
- Phase transformation can be assisted using laser polishing techniques. For instance, the initial $\alpha+\beta$ dual-phase microstructure martensitic phase of TC4 and TC11 was converted into α' martensitic phase.
- Pulsed wave and ultra-short pulse-based polishing have been found to be compelling in cases of selective polishing of surface, micro polishing with less power consumption.
- Laser shock peening techniques have effectively reduced residual tensile stress and generated favourable compressive stress in the material. A refined structure with increased fatigue strength and enhanced resistance to wear can be obtained when post-processed using laser shock peening.
- It can be concluded that laser finishing techniques have great potential in polishing additive manufactured components. However, a few challenges, such as oxidation and carbonization of the workpiece, waviness and ripples formed during polishing and effective removal of track marks from the initially obtained AM samples, should be worked upon to increase the applicability of laser-based polishing methods.

Table 8.4 Materials roughness after laser processing

List of materials with the type of process manufactured	Type of laser post processing	Initial roughness	Final roughness
Ti6al4V (SLA) [47]	Continuous Wave Laser Polishing	12	1.5
Ti6Al4V (SLM) [17]	Pulsed Wave Laser Polishing	0.197	0.096
Ti6Al4V (SLM) [35]	Nano Pulsed Laser polishing	5.226	0.375
Ti-6.5Al-3.5Mo-1.5Zr-0.3Si (SLM) [35]	Nano Pulsed Laser polishing	7.21	0.73
Ti6Al4V (SLM) [44]	Continuous Wave Laser Polishing	3.6	2.36
Ti6Al4V (SLM) [48]	Continuous Wave Laser Polishing	10.2	2.4
CoCr alloy (SLM) [38]	Pulsed Wave Laser Polishing	4.65	0.28
Aluminum Fiber/Polylactide Acid (Al/PLA) Composite (FDM) [29]	Pulsed Wave Laser Polishing	5.64	0.32
ABS (FDM) [51]	Laser-Assisted Post Processing	15.55	10.596
CoCr Alloy (SLM) [39]	Pulsed Wave Laser Polishing	5	0.45
Inconel 625 (SLM) [41]	Continuous Wave Laser Polishing	10.04	0.79
18 Maraging 300 Steel (SLM) [31]	Laser Ablation	3.3	19.7
Stainless Steel (SS316L) [33]	Pulsed Wave Laser Polishing	3.8	0.2
Sintered Stainless Steel and Infiltrated Bronze (SLS) [49]	Pulsed Wave Laser Polishing	7.8	1.49
AISI 316L Stainless Steel Powder (SLM) [50, 53]	Continuous Wave Laser Polishing	12	1.5
SS316L (3DPrint) [33]	Pulsed Wave Laser Polishing	3.8	0.2
AlSi10Mg (SLM) [32]	Continuous Wave Laser Polishing	8.7	0.2
X2CrNiMo17-12-2 Steel (SLM) [32]	Pulsed Wave Laser Polishing	12.5	0.35
Ti6Al4V (SLM) [52]	Nano Pulsed Laser polishing	10.2	2.1
Ti6Al4V (SLM) [52]	Continuous Wave Laser Polishing	6.5	0.3

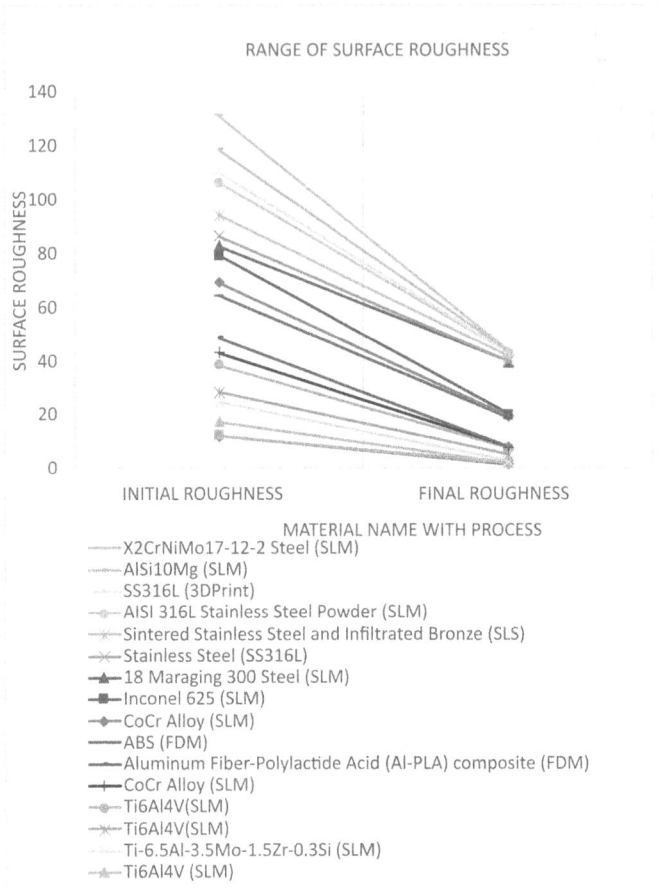

Figure 8.18 Surface roughness vs post-processing for different material.

REFERENCES

[1] Gibson, I., Rosen, D.W., & Stucker, B.E. (2015). *Additive manufacturing technologies: 3D printing, rapid prototyping, and direct digital manufacturing.* Springer.

[2] Kumar, S. (2003). Selective laser sintering: A qualitative and objective approach. *JOM, 55*, 43–47.

[3] Biamino, S., Penna, A., Ackelid, U., Sabbadini, S., Tassa, O., Fino, P., Pavese, M., Gennaro, P., & Badini, C. (2011). Electron beam melting of Ti–48Al–2Cr–2Nb alloy: Microstructure and mechanical properties investigation. *Intermetallics, 19*, 776–781.

[4] Shui, X.J., Yamanaka, K., Mori, M., Nagata, Y., Kurita, K., & Chiba, A. (2017). Effects of post-processing on cyclic fatigue response of a titanium alloy additively manufactured by electron beam melting. *Materials Science and Engineering A-structural Materials Properties Microstructure and Processing, 680*, 239–248.

[5] Maheshwari, S., Siddharth, A., & Alam, Z. (2021). Implementation of cyber-physical systems in additive manufacturing to sustain Covid-19 pandemic. In: Semwal, T. & Iqbal, F., eds. *Cyber-Physical Systems* (1st ed.). CRC Press, Boca Raton.

[6] Jetting, B. (2015). *Additive Manufacturing Technologies: 3-D Printing, Rapid Prototyping, and Direct Digital Manufacturing.* Gibson, I., Rosen, D.W., & Stucker, B.E., eds. Springer, New York.

[7] Ke, W., Oliveira, J., Cong, B., Ao, S., Qi, Z., Peng, B., & Zeng, Z. (2021). Multi-layer deposition mechanism in ultra-high-frequency pulsed wire arc additive manufacturing (WAAM) of NiTi shape memory alloys. *Additive Manufacturing, 50,* 102513.

[8] Zhou, Y., Lin, X., Kang, N., Huang, W., & Wang, Z. (2020). Mechanical properties and precipitation behavior of the heat-treated wire + arc additively manufactured 2219 aluminium alloy. *Materials Characterization, 110735.*

[9] Chang, T., Fang, X., Liu, G., Zhang, H., & Huang, K. (2022). Wire and arc additive manufacturing of dissimilar 2319 and 5B06 aluminum alloys. *Journal of Materials Science & Technology, 124,* 65–75.

[10] Willenborg, E. Polishing with laser radiation. In: Poprawe, R., ed. *Tailored Light 2.* Berlin: Springer, 2011, 196–203.

[11] Vadali, M., Ma, C., Duffie, N.A., Li, X., & Pfefferkorn, F.E. (2012). Effects of laser pulse duration on pulse laser micro polishing. *J. Micro Nano-Manuf, 1*(1), 011006.

[12] Morrow, J.D., Wang, Q., Duffie, N.A., & Pfefferkorn, F.E. (2014). Effects of Pulsed Laser Micro Polishing on Microstructure and Mechanical Properties of S 7 Tool Steel ICOMM 2014 No.103.

[13] Simões, J., Riva, R., & Miyakawa, W. (2018). High-speed Laser-Induced Periodic Surface Structures (LIPSS) generation on stainless steel surface using a nanosecond pulsed laser. *Surface and Coatings Technology, 344,* 423–432.

[14] Perry, T.L., Werschmoeller, D., Li, X., Pfefferkorn, F.E., & Duffie, N.A. (2009). The effect of laser pulse duration and feed rate on pulsed laser polishing of microfabricated nickel samples. *Journal of Manufacturing Science and Engineering-transactions of The Asme, 131,* 031002.

[15] Nüsser, C., Sändker, H., & Willenborg, E. (2013). Pulsed laser micro polishing of metals using dual-beam technology. *Physics Procedia, 41,* 346–355.

[16] Perry, T.L., Werschmoeller, D., Li, X., Pfefferkorn, F.E., & Duffie, N.A. (2009). Pulsed laser polishing of micro-milled Ti6Al4V samples. *Journal of Manufacturing Processes, 11,* 74–81.

[17] Pfefferkorn, F.E., Duffie, N.A., Li, X., Vadali, M., & Ma, C. (2013). Improving surface finish in pulsed laser micro polishing using thermocapillary flow. *Cirp Annals-manufacturing Technology, 62,* 203–206.

[18] Gaudet, S., Detavernier, C., Kellock, A., Desjardins, P., & Lavoie, C. (2006). Thin film reaction of transition metals with germanium. *Journal of Vacuum Science and Technology, 24,* 474–485.

[19] Ukar, E., Lamikiz, A., Lacalle, L.N., Pozo, D.D., & Arana, J.L. (2010). Laser polishing of tool steel with CO2 laser and high-power diode laser. *International Journal of Machine Tools & Manufacture, 50,* 115–125.

[20] Mingareev, I., Bonhoff, T., El-Sherif, A.F., & Richardson, M.C. (2013). Femtosecond laser post-processing of metal parts produced by laser additive manufacturing. *Journal of Laser Applications, 25*(5): 052009.

[21] Perry, T.L., Werschmoeller, D., Duffie, N.A., Li, X., & Pfefferkorn, F.E. (2009). Examination of selective pulsed laser micropolishing on microfabricated nickel samples using spatial frequency analysis. *Journal of Manufacturing Science and Engineering-transactions of The Asme*, *131*, 021002.

[22] Bordatchev, E.V., Hafiz, A.M., & Tutunea-Fatan, O.R. (2014). Performance of laser polishing in finishing of metallic surfaces. *The International Journal of Advanced Manufacturing Technology*, *73*, 35–52.

[23] Taylor, L.L., Xu, J., Pomerantz, M., Smith, T.R., Lambropoulos, J.C., & Qiao, J. (2019). Femtosecond laser polishing of germanium [Invited]. *Optical Materials Express*, *9*, 4165–4177.

[24] Ruck, S., Harrison, D.K., Silva, A.K., Kleefoot, M., & Riegel, H. (2021). *Single Track* ultra-short pulsed laser ablation on additive manufactured metallic surfaces. *Journal of Laser Micro/Nanoengineering*, *16*, 36–41.

[25] Ramos, J.A., Bourell, D.L., & Beaman, J.J. (2002). Surface over-melt during laser polishing of indirect-SLS metal parts. *MRS Proceedings*, *758*, 53–61.

[26] Morawetz, K., Trinschek, S., & Gurevich, E.L. (2022). Interplay of viscosity and surface tension for ripple formation by laser melting. *Physical Review B*, *105*, 035415.

[27] Temmler, A., Cortina, M., Ross, I., Küpper, M.E., & Rittinghaus, S. (2021). Laser Micro Polishing of Tool Steel 1.2379 (AISI D2): Influence of intensity distribution, laser beam size, and fluence on surface roughness and area rate. *Metals*.

[28] Karthik, D., Kalainathan, S., & Swaroop, S. (2015). Surface modification of 17-4 PH stainless steel by laser peening without protective coating process. *Surface & Coatings Technology*, *278*, 138–145.

[29] Chen, L., & Zhang, X. (2019). Modification the surface quality and mechanical properties by laser polishing of Al/PLA part manufactured by fused deposition modeling. *Applied Surface Science*, *492*, 765–775.

[30] Hofele, M., Roth, A., Hegele, P., Schubert, T., Schanz, J., Harrison, D.K., De Silva, A.K., & Riegel, H. (2022). Influence of laser polishing on the material properties of aluminium L-PBF components. *Metals*, *12*, 5–750.

[31] Solheid, J.D., Seifert, H.J., & Pfleging, W. (2018). *Laser-Assisted Post-Processing of Additive Manufactured Metallic Parts*.

[32] Hofele, M., Schanz, J., Burzic, B., Lutz, S., Merkel, M., & Riegel, H. (2017). *Laser Based Post Processing of Additive Manufactured Metal Parts*.

[33] Bhaduri, D., Penchev, P., Batal, A., Dimov, S.S., Soo, S.L., Sten, S., Harrysson, U., Zhang, Z., & Dong, H. (2017). Laser polishing of 3D printed mesoscale components. *Applied Surface Science*, *405*, 29–46.

[34] Tian, Y., Chen, C., Li, S., & Huo, Q.H. (2005). Research progress on laser surface modification of titanium alloys. *Applied Surface Science*, *242*, 177–184.

[35] Ma, C.P., Guan, Y., & Zhou, W. (2017). Laser polishing of additive manufactured Ti alloys. *Optics and Lasers in Engineering*, *93*, 171–177.

[36] Marimuthu, S., Eghlio, R.M., Pinkerton, A., & Li, L. (2013). Coupled computational fluid dynamic and finite element multiphase modeling of laser weld bead geometry formation and joint strengths. *Journal of Manufacturing Science and Engineering-Transactions of the Asme*, *135*, 011004.

[37] Góra, W.S., Tian, Y., Cabo, A.P., Ardron, M., Maier, R.R., Prangnell, P.B., Weston, N.J., & Hand, D.P. (2016). Enhancing surface finish of additively manufactured Titanium and Cobalt Chrome elements using laser based finishing. *Physics Procedia*, *83*, 258–263.

[38] Yung, K.C., Xiao, T.Y., Choy, H.S., Wang, W.J., & Cai, Z. (2018). Laser polishing of additive manufactured CoCr alloy components with complex surface geometry. *Journal of Materials Processing Technology*, 262, 53–64.

[39] Wang, W.J., Yung, K.C., Choy, H.S., Xiao, T.Y., & Cai, Z. (2018). Effects of laser polishing on surface microstructure and corrosion resistance of additive manufactured CoCr alloys. *Applied Surface Science*.

[40] Tian, Y., Góra, W.S., Cabo, A.P., Parimi, L.L., Hand, D.P., Tammas-Williams, S., & Prangnell, P.B. (2018). Material interactions in laser polishing powder bed additive manufactured Ti6Al4V components. *Additive Manufacturing*, 20, 11–22.

[41] Jinoop, A.N., Subbu, S.K., Paul, C.P., & Palani, I.A. (2019). Post-processing of Laser Additive Manufactured Inconel 718 using laser shock peening. *International Journal of Precision Engineering and Manufacturing*, 20, 1621–1628.

[42] Luo, S., He, W., Chen, K., Nie, X., Zhou, L., & Li, Y. (2018). Regain the fatigue strength of laser additive manufactured Ti alloy via laser shock peening. *Journal of Alloys and Compounds*, 750, 626–635.

[43] Jiang, Q., Li, S.B., Zhou, C., Zhang, B., & Zhang, Y. (2021). Effects of laser shock peening on the ultra-high cycle fatigue performance of additively manufactured Ti6Al4V alloy. *Optics and Laser Technology*, 144, 107391.

[44] Xinyuan, J., Lan, L., Shuang, G., He, B., & Rong, Y. (2020). Effects of laser shock peening on microstructure and fatigue behavior of Ti–6Al–4V alloy fabricated via electron beam melting. *Materials Science and Engineering A-structural Materials Properties Microstructure and Processing*, 780, 139199.

[45] Chi, J., Cai, Z., Wan, Z., Zhang, H., Chen, Z., Li, L., Li, Y., Peng, P., & Guo, W. (2020). Effects of heat treatment combined with laser shock peening on wire and arc additive manufactured Ti17 titanium alloy: Microstructures, residual stress and mechanical properties. *Surface & Coatings Technology*, 396, 125908.

[46] Kalentics, N., Boillat, E., Peyre, P., Ćirić-Kostić, S., Bogojević, N., & Logé, R.E. (2017). Tailoring residual stress profile of Selective Laser Melted parts by Laser Shock Peening. *Additive manufacturing*, 16, 90–97.

[47] Vaithilingam, J., Goodridge, R.D., Hague, R.J., Christie, S.D., & Edmondson, S. (2016). The effect of laser remelting on the surface chemistry of Ti6al4V components fabricated by selective laser melting. *Journal of Materials Processing Technology*, 232, 1–8.

[48] Marimuthu, S., Triantaphyllou, A., Antar, M., Wimpenny, D.I., Morton, H., & Beard, M.A. (2015). Laser polishing of selective laser melted components. *International Journal of Machine Tools & Manufacture*, 95, 97–104.

[49] Lamikiz, A., Sánchez, J.A., Lacalle, L.N., & Arana, J.L. (2007). Laser polishing of parts built up by selective laser sintering. *International Journal of Machine Tools & Manufacture*, 47, 2040–2050.

[50] Yasa, E., Kruth, J.P., & Deckers, J.H. (2011). Manufacturing by combining selective laser melting and selective laser erosion/laser re-melting. *Cirp Annals-manufacturing Technology*, 60, 263–266.

[51] Taufik, M., & Jain, P.K. (2017). Laser assisted finishing process for improved surface finish of fused deposition modelled parts. *Journal of Manufacturing Processes*, 30, 161–177.

[52] Li, J., & Zuo, D.W. (2021). Laser polishing of additive manufactured Ti6Al4V alloy: A review. *Optical Engineering*, 60, 020901–020902.

Index

For Product Safety Concerns and Information please contact our EU
representative GPSR@taylorandfrancis.com
Taylor & Francis Verlag GmbH, Kaufingerstraße 24, 80331 München, Germany

www.ingramcontent.com/pod-product-compliance
Lightning Source LLC
Chambersburg PA
CBHW060355220326
41598CB00023B/2924

9 781032 616018